AF745459

Issues on Risk Analysis for Critical Infrastructure Protection

Edited by Vittorio Rosato and Antonio Di Pietro

Published in London, United Kingdom

IntechOpen

Supporting open minds since 2005

Issues on Risk Analysis for Critical Infrastructure Protection
http://dx.doi.org/10.5772/intechopen.91097
Edited by Vittorio Rosato and Antonio Di Pietro

Contributors
Aleksandar S. Jovanovic, Marjan Jelic, Somik Chakravarty, Esra Tepeli, Luisa Franchina, Alessandro Calabrese, Giulia Inzerilli, Enrico Scatto, Marieke Klaver, Eric Luiijf, Chiara Foglietta, Stefano Panzieri, Ugonna Nkwunonwo, Andrea Carpignano, Raffaella Gerboni, Daniele Grosso, Andrea Bologna, Vittorio Rosato, Yiorgos Stergiopoulos, Giulio Smedile, Kotzanikolaou Panagiotis, Antonio Di Pietro, David Rehak, Martin Hromada, Anna Gazi, Ilias Gkotsis, Evita Agrafioti, Anastasia Chalkidou, Fabio Bolletta, Clemente Fuggini, Karolina Jurkiewicz, Gabriele Oliva, Giacomo Assenza, Luca Faramondi, Roberto Setola, Ernesto Del Prete, Fabio Pera, Anatoly Lepikhin, Victor Leschenko, Nikolay Machutov, Marica Mazurek

© The Editor(s) and the Author(s) 2021
The rights of the editor(s) and the author(s) have been asserted in accordance with the Copyright, Designs and Patents Act 1988. All rights to the book as a whole are reserved by INTECHOPEN LIMITED. The book as a whole (compilation) cannot be reproduced, distributed or used for commercial or non-commercial purposes without INTECHOPEN LIMITED's written permission. Enquiries concerning the use of the book should be directed to INTECHOPEN LIMITED rights and permissions department (permissions@intechopen.com).
Violations are liable to prosecution under the governing Copyright Law.

Individual chapters of this publication are distributed under the terms of the Creative Commons Attribution 3.0 Unported License which permits commercial use, distribution and reproduction of the individual chapters, provided the original author(s) and source publication are appropriately acknowledged. If so indicated, certain images may not be included under the Creative Commons license. In such cases users will need to obtain permission from the license holder to reproduce the material. More details and guidelines concerning content reuse and adaptation can be found at http://www.intechopen.com/copyright-policy.html.

Notice
Statements and opinions expressed in the chapters are these of the individual contributors and not necessarily those of the editors or publisher. No responsibility is accepted for the accuracy of information contained in the published chapters. The publisher assumes no responsibility for any damage or injury to persons or property arising out of the use of any materials, instructions, methods or ideas contained in the book.

First published in London, United Kingdom, 2021 by IntechOpen
IntechOpen is the global imprint of INTECHOPEN LIMITED, registered in England and Wales, registration number: 11086078, 5 Princes Gate Court, London, SW7 2QJ, United Kingdom
Printed in Croatia

British Library Cataloguing-in-Publication Data
A catalogue record for this book is available from the British Library

Additional hard and PDF copies can be obtained from orders@intechopen.com

Issues on Risk Analysis for Critical Infrastructure Protection
Edited by Vittorio Rosato and Antonio Di Pietro
p. cm.
Print ISBN 978-1-83962-620-3
Online ISBN 978-1-83962-621-0
eBook (PDF) ISBN 978-1-83962-628-9

For EU product safety concerns:
IN TECH d.o.o., Prolaz Marije Krucifikse Kozulić 3, 51000 Rijeka, Croatia, info@intechopen.com or visit our website at intechopen.com.

We are IntechOpen, the world's leading publisher of Open Access books

Built by scientists, for scientists

5,300+
Open access books available

131,000+
International authors and editors

155M+
Downloads

156
Countries delivered to

Our authors are among the
Top 1%
most cited scientists

12.2%
Contributors from top 500 universities

CLARIVATE ANALYTICS BOOK CITATION INDEX INDEXED

WEB OF SCIENCE™

Selection of our books indexed in the Book Citation Index
in Web of Science™ Core Collection (BKCI)

Interested in publishing with us?
Contact book.department@intechopen.com

Numbers displayed above are based on latest data collected.
For more information visit www.intechopen.com

Meet the editors

Vittorio Rosato has a laurea degree in Physics (University of Pisa, 1979) and a Ph.D. in Physics (University of Nancy, 1986). He was a research associate at the University College of Wales in Aberystwyth (UK) and EURATOM Fellow at the CEA Centre de Recherche Nucleaires de Saclay (F). His research field has been computational methods in condensed matter physics and complexity science. He is currently the research director at ENEA Casaccia Research Centre, at the Laboratory of Analysis and Protection of Critical Infrastructures, and at the Italian Node of the European Infrastructure Simulation and Analysis Centre (EISAC.it). His current research activities span from risk analysis to the design of decision support systems for the management of complex technological networks.

Antonio Di Pietro earned his master's degree in Informatics Engineering at the Sapienza University of Rome in 2004 and his Ph.D. in the same field at Roma Tre University in 2015. He has been working as a researcher at ENEA since 2007. His current research activities include modeling and simulation of critical infrastructures and the development of decision support systems integrating natural hazard modeling. He took part in several European and Italian national research projects and acted as an adviser in some evaluation studies commissioned by the EU in the field of critical infrastructure protection. He has been an adviser of several MSc students and also a professor in software engineering, programming languages, databases, and distributed applications courses.

Contents

Preface

Critical infrastructure (CI) provides, better than any other metaphor, the representation of a large "System of Systems" providing primary functions and vital services for societal life. They support citizens' activities and constitute a necessary component for all industrial and economical value chains. Their protection and the enhancement of systemic resilience must be thus a primary concern of modern countries. CI, together with technological functions and essential services, carries on a strong social value as it transports psychological side images related to the perception of public security, social cohesion, and technological efficiency, in a way that the safeguard and optimal management are elemental in contributing to the citizens' trust in public institutions.

Such multiple relevances are, however, accompanied by a number of issues that make CI management and protection difficult, leaving them prone to their intrinsic and extrinsic vulnerabilities of natural and anthropic events that continuously threaten their integrity; for example, voluntary attacks to the physical and the cyber scale are threats. This further increases their complexity leading to the need for identifying new strategies to overcome limitations and achieve their "smart management."

The major issues enhancing complexity and vulnerability in the CI domain are related on the one hand to their mutual interdependencies and, on the other hand, to the current *linearization* of their management. Dependencies and inter-dependencies are due to the intense exchange of services among CI with the consequent emergence of dangerous perturbations that can propagate from one system to other connected systems. Perturbations might expand instabilities, reduce functionality in time and space, and consequently transform a local impact into consequences that might extend on larger scales and last for longer periods of time.

Management *linearization* results from the current ownership fragmentation of CI: different operators own and manage their CI independently from the others (even from those that provide services to them) as if the bundle of CI was only weakly interacting. Only in this case would *linearized* management be effective; however, this is not the case and the strong coupling between CI operators makes the *linearized* management strategy much less effective and unable to produce optimal results, particularly in the case of strong perturbations occurring in extended crises.

In order to overcome the negative effects produced by these issues, technology must provide new tools and new ideas for smarter management of CI that, although accounting for the unavoidable constraints (i.e., ownership fragmentation, industrial competitions among players insisting on the same CI, etc.), can improve the current management efficacy and enhance the resilience at the "systemic" scale. Resilience, by far, is the more important endpoint of all efforts: after having abandoned the unrealistic claim of enduring complete invulnerability to the assets, resilience offers a smart, adaptive property that allows a system to regain its equilibrium configuration, rapidly and effectively, after a reduction (or even the loss of it) due to some perturbations.

Although providing direct services on their own (electricity and other energy products, telecommunication for voice and data, water, and distribution of other products, etc.), specific added-value combinations of CI do allow to realize and dispatch a number of other services (logistics, financial and public health services, etc.); these services, on the one hand, increase the CI relevance as they essentially embrace all domains of a citizen's life but, on the other hand, increases their "attack surface" (i.e., the functions and the points from where they can be hit and receive perturbations).

This volume spans over several areas and highlights a number of different issues related to the management and the protection of CI.

As resilience is the most relevant property, it is described as being enhanced by improving capabilities in the modeling and simulation approach (Foglietta and Panzieri), by improving risk analysis in infrastructure projects (Tepeli), by introducing a criticality index to estimate the economic damage associated to all the hazards (Gerboni et al.), by improving situational awareness (Jovanovic et al.), and by stressing the importance of integrating dependency mechanisms linking different infrastructures in a unique system of systems (Rosato et al). All efforts should be addressed to improve the survivability of critical elements (Oliva et al.); this book describes the attempts of simulating networks, such as gas pipelines, at the specific infrastructure scales (Rehak et al.), and also their subsea installation (Lepikhim et al.). Several contributions deal with the different classes of hazards that menace the physical and control integrity of the assets: the case of threats coming from the cyber domain (Klaver and Luiijf) and other cases coming from flooding and similar natural events that could either intensify damage due to climatic changes or hit underdeveloped countries (Nkwunonwo). Critical infrastructure might impact tourism (Mazurekova) and other activities, as it occurred in 2020, the year of the worldwide pandemic (its report in Italy has been described by Inzerilli et al.).

A major outcome of the last ten years' activities in the research domain of critical infrastructure protection is the understanding of the need for a coherent action at the level of large, international communities, for instance at the level of the European Union (EU), to elaborate a homogeneous level of protection to the most critical infrastructure. The system of critical infrastructure has assumed a transnational identity and cannot be treated anymore as if it were composed of separate entities with local (i.e., at the level of a single country) dimension. To this end, the EU is preparing a new directive (which will be issued in late 2021/early 2022) with the aim of supporting member states to more proactive management of their strategic assets and to a more homogeneous protection level (particularly against the cyber risk and toward new threats that are going to arise due to climatic change).

This new EU initiative calls, with renewed strength, to the establishment of some EU-wide initiative allowing to cope with such new ambitious, albeit unavoidable, goals. A relevant and appropriate suggestion has been proposed, in a far-sighted way, by EU FP7 Project CIPRnet in 2014 (ciprnet.eu): the establishment of a constellation of National Competence Centers devoted to the support of CI operators and public authorities dealing with critical infrastructure protection. The initiative called the European Infrastructure Simulation and Analysis Centre (EISAC) has been boosted by advanced technological systems enabling EISAC centers to provide the needed support: the overviewing of the system of interconnected critical infrastructure through continuous monitoring and 24/7 risk analysis, providing a shareable systemic awareness to all operators to be used to manage them in ordinary times and to recover their functions after a crisis by allowing the establishment of a global optimum configuration rather than of a sequence of local optima. The establishment of the first node of the constellation, EISAC.it (the Italian node), is underway. This will also nucleate and support the birth of new EISAC centers in the other EU member states in the coming years.

These types of initiatives have a two-fold beneficial impact: on the one hand, they can gather new technological platforms and instruments from the R&D domains and provide them an operational endpoint that could effectively provide a significant benefit to citizens. On the other hand, they carry on the idea of better collaboration and cooperation among the different CI owners by adopting a systemic perspective for the protection of critical assets. We must all collaborate on the well-being and the progress of societies. There should be no further return to a "divide and rule" strategy. It should no longer be allowed and must be replaced by a new cooperative model. This is the only appropriate management strategy to manage strategic assets within modern, entangled structures.

Vittorio Rosato and Antonio Di Pietro
ENEA Casaccia Research Centre,
Department of Energy Technologies and Renewable Sources,
Rome, Italy

Section 1

Resilience as a Key Property of Critical Infrastructure

Chapter 1

Resilience in Critical Infrastructures: The Role of Modelling and Simulation

Chiara Foglietta and Stefano Panzieri

Abstract

Resilience and risk are fundamental concepts for critical infrastructure protection, but it is complex to assess them. Modelling critical infrastructure interdependency helps in evaluating the resilience and risk metrics. We propose the MHR approach as a road-map to model infrastructures and it is implemented using CISIApro 2.0. MHR suggests considering three different layers in each infrastructure: holistic, service and reductionist agents. In this chapter, this framework has been tested in a scenario made of a modern telecommunication network, a hospital ward and a smart factory. The scenario takes into account cyber attacks and their consequences on the components, services and holistic nodes. The proposed framework is under validation within the EU H2020 RESISTO project with good results and in various test-beds.

Keywords: resilience metric, risk management, critical infrastructure modelling, simulation

1. Introduction

Critical Infrastructure is an evolving concept. Critical infrastructure was linked to aging public works in the 1980s: the National Council on Public Works Improvement in 1988 focused on public sector infrastructure. In the 1990s, infrastructure was redefined in terms of national security as a consequence of increased international terrorism. The number of critical infrastructure sectors in the National Infrastructure Protection Plan [1] has been enlarged to 17 since 9/11: it includes agriculture and food systems, the defense-industrial base, electricity systems, public health and health care facilities, national monuments, banking and financial systems, drinking water systems, chemical services, commercial buildings, dams, emergency services, nuclear power plants, information technology networks, telecommunications systems, postal and shipping services, transportation systems, and government facilities. Critical infrastructure is identified in Europe under the term "essential services" [2].

Shifting the concept of critical infrastructures has led to more flexibility and adaptability. The sophistication of an already complicated field, on the other hand, is increased, creating more confusion and more doubts. The definition of "lifeline system", [3] was then established by some researchers to assess the efficiency of large, geographically distributed networks during crises caused by adverse events, such as natural disasters or cyber-attacks. Lifelines are classified into six major systems: electricity, gas and liquid fuels, telecommunications, transportation, waste management, and water provision. The economic well-being, security, and protection of our lives are closely related to those systems. Thinking of critical infrastructure across the sub-set of lifelines helps to simplify features common to important support structures and to enhance the performance of large networks, offering visibility into the technical challenges.

Lifeline systems, mostly on the basis of physical proximity and operational interaction, are interdependent. Cables and pipes are placed alongside each other in crowded area, resulting in an elevated risk due to proximity. Damage to one infrastructure component, such as an electrical cable, will easily ripple into damage to adjacent components, such as telecommunications cables and gas mains, with system-wide implications.

Lifeline systems are dependent on each other. Electric power networks, for example, supply electricity for pumping stations, storage facilities, and equipment control for transmission and distribution systems for oil and natural gas. Oil provides fuel and lubricants for generators, and natural gas provides energy for generating stations, compressors, and storage, all of which are required for the operation of electric power networks.

In the Merriam-Webster Dictionary, resilience is defined as "the capability of a strained body to recover its size and shape after deformation caused especially by compressive stress" [4]. Definitions vary slight, but all of them relate the principle of resilience to physical stress recovery.

A notable change from securing critical infrastructures to ensuring that communities are resilient has taken place following Hurricane Katrina. Furthermore, the concept of resilience is evolving, as the idea of critical infrastructures. In its present form, a society's resilience is an overarching attribute that reflects the degree of community preparedness and the ability to respond to a crisis and rebound from it. Since lifelines are intimately linked to the economic well-being, security, and social fabric of a community, community resilience is closely related to the initial strength and gradual recovery of lifelines.

Debate over the concept of resilience is likely to persist, and refinements and elaborations of the term are to be expected. A framework for defining resilience has been suggested by the Multidisciplinary Center for Earthquake Engineering Research (MCEER) [5]. Resilience for both physical and social systems can be conceptualized as having four infrastructural qualities:

- Robustness: the inherent strength or resistance in a system to withstand external demands without degradation or loss of functionality.

- Redundancy: system property that under stress allows for alternate solutions, decisions, and substitutions.

- Resourcefulness: the capacity to coordinate needed assets and services in crises.

- Rapidity: the speed at which disruption can be overcome and safety, services, and financial stability restored.

As shown in **Figure 1**, an infrastructural performance, such as robustness, $Q(t)$, can be visualized as a percentage that varies with time. For buildings, $Q(t)$ may be the percentage of structural or functional integrity. For lifelines, $Q(t)$ may be the percentage of customers that successfully receive power or drinking water. Prior to a natural hazard, severe accident, terrorist act, or a general disruption, $Q(t)$ is at 100

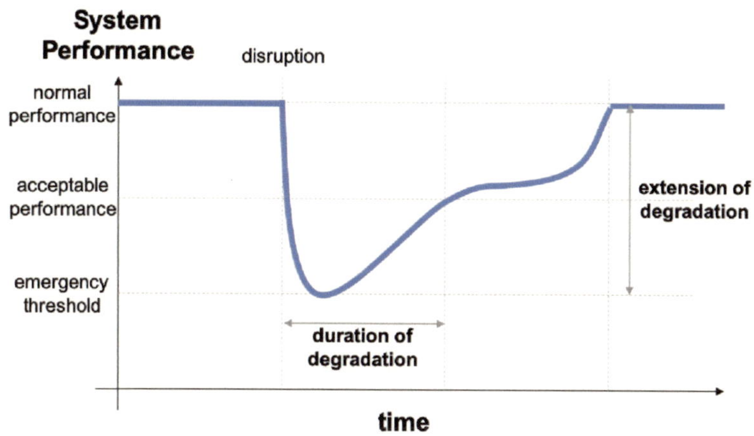

Figure 1.
The resilience profile.

percent; in picture is defined as normal performance. If the system is fully robust, it remains at 100 percent even during disruptions. Total loss of service results in 0 percent of $Q(t)$. If system disturbance occurs at time t_0, in response to, for example, an earthquake or hurricane, damage to the infrastructure may reduce the performance to less than 100 percent, the emergency threshold. Level of service, as reflected by the robustness of the system, is a function of the probability and consequences of damage. Robustness is restored over time; at time t_1, the system is returned to its original capacity. We called "duration of degradation" the time for the system to bounce back to an acceptable performance.

For a community or an infrastructure, the loss of resilience, R, can be measured as the expected loss in quality (probability of failure) over the time to recovery, $t_1 - t_0$. Thus, mathematically, R is defined as:

$$R = \int_{t_0}^{t_1} Q(t)dt \tag{1}$$

The resilience indicator, R, is a simple measure for quantifying resilience. In [5], additional mathematical developments of this notion cover the probabilistic and multidimensional aspects of resilience.

1.1 Contributions

The modeling method used in this chapter is based on the methodology of Mixed Holistic Reductionist (MHR), where each infrastructure is divided into components (reductionist layer), services (service layer) and holistic nodes (holistic layer). The MHR approach is a guideline on how we can decompose each infrastructure and how we can define the interconnection among the different components. It also allows the identification of the right abstraction level due to the available information.

The agent-based simulator, called CISIApro 2.0, is then used to implement this approach. This simulator presents the consequences of adverse and positive events in an interdependent scenario. In real-time, this simulator runs connected to a SCADA (Supervisory Control And Data Acquisition) control center to receive current information on faults and linked to an Intrusion Detection System (IDS) to acquire actual threats and on-going cyber-attacks. CISIApro 2.0 integrates heterogeneous data to improve the situational awareness of operators and their decision-making process. This version of the simulator has been improved considering the telecommunication features. Specifically they are:

- Elements with multiple services
- Dynamic links
- Routing links
- Propagation models for ring topologies
- Continuous and discrete dynamics simulation inside the agents
- The possibility of co-simulating external dynamics
- The ability of revoke services

1.2 Organizations

This chapter is composed of the following sections: Section 2 analyses the idea of risk and resilience; Section 3 reviews the literature on critical infrastructures simulator; Section 4 presents the MHR approach while the simulator CISIApro 2.0 is described in Section 5; a telecommunication case study is summarised in Section 6; conclusions and future works are in Section 7.

2. The concepts of risk and resilience

The concepts of risk and resilience are similar and generally closely linked: improving the system's resilience requires reducing risk. Risk is commonly structured in terms of preparedness, mitigation measures, reaction capabilities, and recovery processes; anticipation, absorption, adaptation and recovery are the typical components of resilience.

Owners and operators can improve the resilience of critical infrastructures by specific operations: withstanding specific threats, reducing or mitigating potential impacts, returning to normal operations if such degradation occurs. A resilience methodology includes increasing preparedness for an incident, implementing redundancy to mitigate the effects of an incident, and strengthening the coordination and execution of response and recovery procedures, for emergency action and business continuity.

There are five main steps in the resilience cycle: prepare, prevent, protect, response and recover. The resilience cycle must consider the consequences of interdependencies among critical infrastructures. The tool we present in this chapter, called CISIApro 2.0, aims to assess the consequences of adverse events on critical infrastructures in terms of components, services and also holistic agents. CISIApro 2.0 usually helps the operators in the recovery phase, knowing which are the possible consequences of actual adverse events.

The Department of Homeland Security (DHS) defines risk as "the potential for an unwanted outcome resulting from an incident, event, or occurrence, as determined by its likelihood and the associated consequences" [6]. Thus, risk is historically characterized as a function of three elements: the threats to which an asset is susceptible, the vulnerabilities of the asset to the threat, and the consequences potentially generated by the asset's deterioration.

Threat is a "natural or man-made occurrence, individual, entity, or action that has or indicates the potential to harm life, information, operations, the environment, and/or property" [6]. Sometimes the term hazard, which can be defined as a "natural or man-made source or cause of harm or difficulty" [6], is used instead of threat. However, a "hazard differs from a threat in that a threat is directed at an entity, asset, system, network, or geographic area, while a hazard is not directed" [6]. Vulnerability is a "physical feature or operational attribute that renders an entity open to exploitation or susceptible to a given hazard" [6]. Consequences are the "effects of an event, incident, or occurrence" [6].

The challenge is to determine where and how resilience integrates into risk assessment as risk is a feature of threats and hazards, weaknesses, and consequences. Resilience, as defined by DHS, is the "ability to resist, absorb, recover from or successfully adapt to adversity or a change in conditions" [6]. The DHS lexicon also states that "Resilience can be factored into vulnerability and consequence estimates when measuring risk" [6]. Therefore, the resilience will have an effect on both vulnerability and consequences.

On the basis of these characteristics, it is possible to develop specific indicators and metrics to assess the risk to an organization or an infrastructure. Considering a threat or hazard (man-made or natural), the vulnerability and resilience of an organization will impact the potential consequences of an event. The interaction between the elements of risk is complex and made more so when one considers the transfer of risk between assets in the case of a threat by an intelligent adversary.

3. Literature review on modelling interdependency

In literature, three main methodologies for the modelling approaches of critical infrastructure modelling are presented: agent-based simulation, input–output analysis and network modelling. Please refer to [7] for heterogeneous and/or unclassified approaches.

Each infrastructure is considered by agent-based simulations to be a complex adaptive structure, consisting of agents representing single aspects of the infrastructure itself. Different agents can be modelled at different degrees of abstraction based on the proposed level of resolution modelling. The primary benefit of agent-based simulation is the ability to establish synergistic behaviors as agents begin to work together [8].

The second method is based on the economic theory of Input–Output proposed by Leontief in the early 1930s, but later adapted to modelling infrastructures. Haimes and Jiang developed the linear input–output inoperability model (IIM) to research the impact of interdependencies on the inoperability of interconnected networked systems [9]. The key benefit of the IIM and its improvements is that the suggested solution is simple and flexible. IIM is usually confined to the financial costs of interdependencies.

In recent years, researchers have investigated new approaches to interdependency modelling of infrastructures. The most promising technique is based on graph and network theory. This approach uses abstract graphs made of nodes and arcs to describe infrastructures, representing links between components within infrastructures. The key benefit is to leverage closed form expressions and numerical simulations to characterise their topology, performance and uncertainty.

4. Mixed Holistic Reductionist (MHR) approach

In this chapter, we propose an already applied approach, for helping during the modelling phase. To maximize the benefits of holistic and reductionist approaches, the Mixed Holistic Reductionist (MHR) [10] methodology was developed. The key goal of MHR approach is to provide a potential road-map to model critical infrastructures and their interdependencies properly.

In holistic modeling, infrastructures are seen as specific agents with defined boundaries and functional properties, creating a global and overall analysis. The purpose of presenting an infrastructure as a single element is to define the various infrastructures and their geographical extent. The volume of data needed for modeling activities is very limited at this stage and can be found in public data-sets.

In the other hand, to better appreciate the overall infrastructure, the reductionist approach stresses the need to thoroughly understand the roles and behaviours of individual components. The reductionist approach drills down to each component in terms of inputs and outputs. At this level of abstraction is easy to find dependencies between equipment and single components.

Various levels of analysis are required in modelled systems and their boundaries are lost in the event of complex case studies. For the MHR model, either a top-down or bottom-up approach might see relationships between infrastructures at different levels. The other key benefit is to model infrastructures at at multiple complexity levels, taking into account the quantity of data available.

The connection point between the two abstraction levels, i.e. holistic and reductionist approaches, is the quality of services (in the following, abbreviated as "service") which is a key element for operators. This layer describes functional relationships between components and infrastructure at different levels of granularity. Services to clients and to other interconnected infrastructures are specifically treated in MHR as a middle layer between holistic and reductionist agents.

The MHR allows us to reach the right level of detail with minimal data and collected information. Some important considerations can be summarised in the following:

- Each infrastructure is modelled starting from the identification of components and their interactions;
- Each layer is defined with an appropriate level of abstraction based on information coming from end-users, stakeholders and open documents;
- Each component (we called it entity or agent) must be described in a way to decouple it from other components: the behaviour of the component must depend on the valued explicitly exchanged with the other components;
- The simulator must be able to represent any type of agent's behaviour for adapting to the specific reference scenario.

MHR approach allows to define three different typologies of agents: holistic agent, service agent and reductionist agents.

The infrastructure as a whole (or its general organizational divisions) is represented by a holistic agent (**Figure 2**) to provide a model that can understand the global interactions between infrastructures.

A service agent represents a logical or organizational aspect, that provides an aggregate resource as the remote control: the remote control generally provides supervision, by means of software and data collection. Data can be collected through telecommunication network or field equipment in case of a geographically distributed infrastructure. In **Figure 3**, a service component is depicted considering the classical model of an agent in CISIApro 2.0. Some examples of service are: the ability to supply customers, the ability to produce resources, the ability to

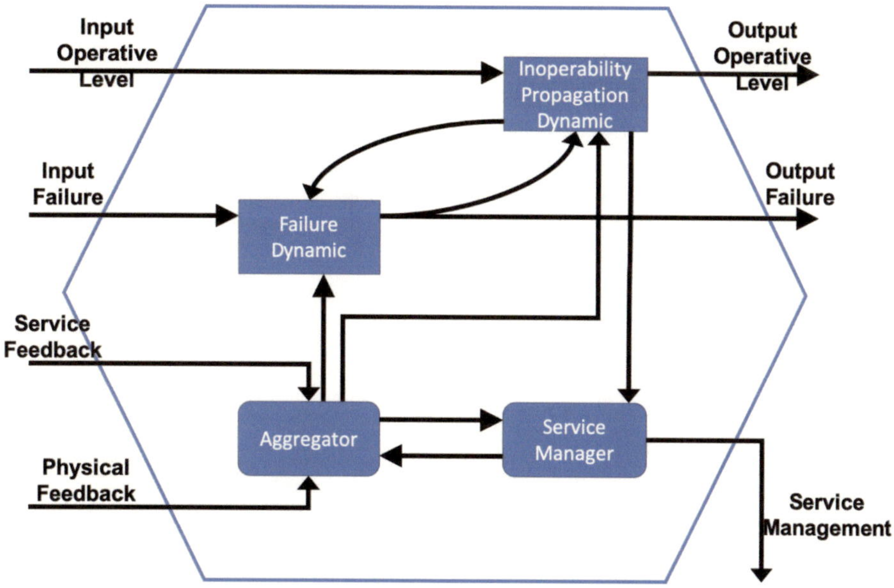

Figure 2.
The holistic agent representation.

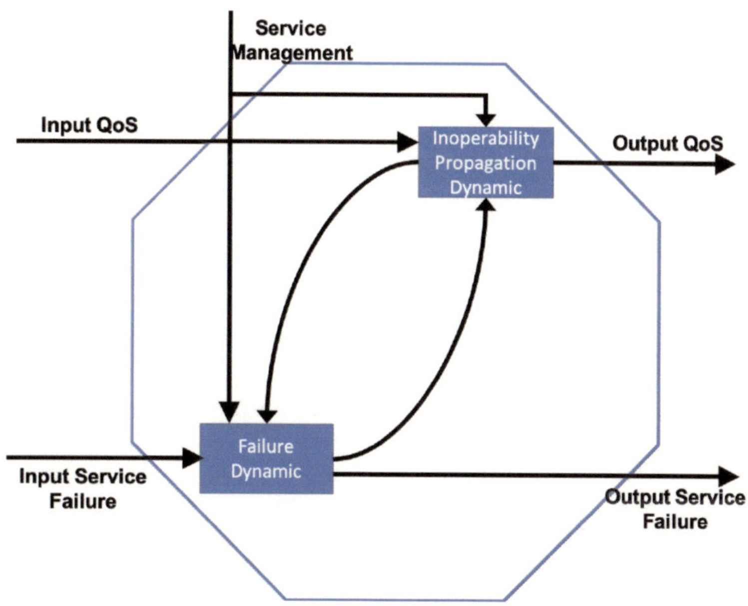

Figure 3.
The service agent representation.

change topology, the aggregate state of a subset of specific and important components.

Finally, with a reductionist agent, we can represent, with the right degree of abstraction, all physical or aggregated entities of the overall system. In **Figure 4**, the representation of a reductionist component is depicted. The picture does not explicitly consider a cyber threat: this malicious event can be represented in the same way as an input failure with a suitable "cyber dynamic".

Finally, we can represent, with the right degree of abstraction, physical or aggregated components of the overall system with a reductionist agent. The

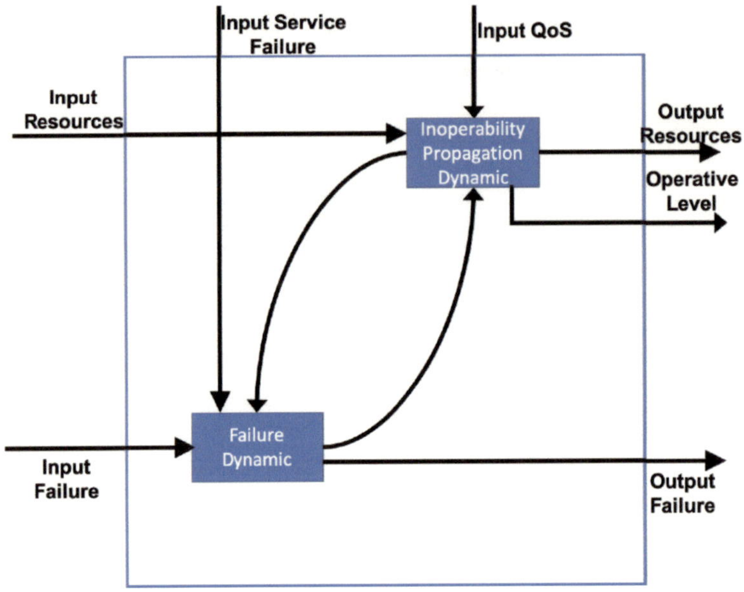

Figure 4.
The reductionist agent representation.

representation of a reductionist aspect is represented in the **Figure 4**. The input failure contains natural disaster events, failures and faults, but also cyber threats.

5. CISIApro 2.0 simulator

In this chapter, CISIApro 2.0 simulates the impact of anomalies and security attacks on the communication infrastructure and on the interlinked CIs. It will also support the decision-making process allowing a "what-if analysis" by simulating the application of countermeasures and reconfiguration and their impact on system resilience.

CISIApro 2.0 (Critical Infrastructure Simulation by Interdependent Agents) [11] is a software engine able to calculate complex cascading effects, taking into account (inter)dependencies and faults propagation among the involved complex systems.

CISIApro 2.0 is an Agent-Based simulation software consisting primarily of two modules, see **Figure 5**. The first one is the off-line tool in which it is possible to design and implement complex and highly interdependent scenarios. While the second one is the on-line tool which is implemented in Simulink (Mathworks).

CISIApro 2.0 is a database-centric architecture in which the database plays a key role as deonstrated in **Figure 5**. This implies a centralized asynchronous design that allows good modularity and scalability where each part of the IT infrastructure interacts, independently, with the centralized database in order to access the last data from the field (e.g. SCADA Systems), Complex Event Processing and generic IoT (Internet of Things) data systems, but also the simulation's outputs.

Using the Mixed-Holistic-Reductionist (MHR) approach, modelling complex interdependent systems is a prerequisite to produce an effective model. Once modelled the involved scenario, with MHR methodology can be applied with CISIApro 2.0.

From this point of view, CISIApro 2.0 engine does not only analyze actual situation and calculate the risk projected in the possible near future but, first, it

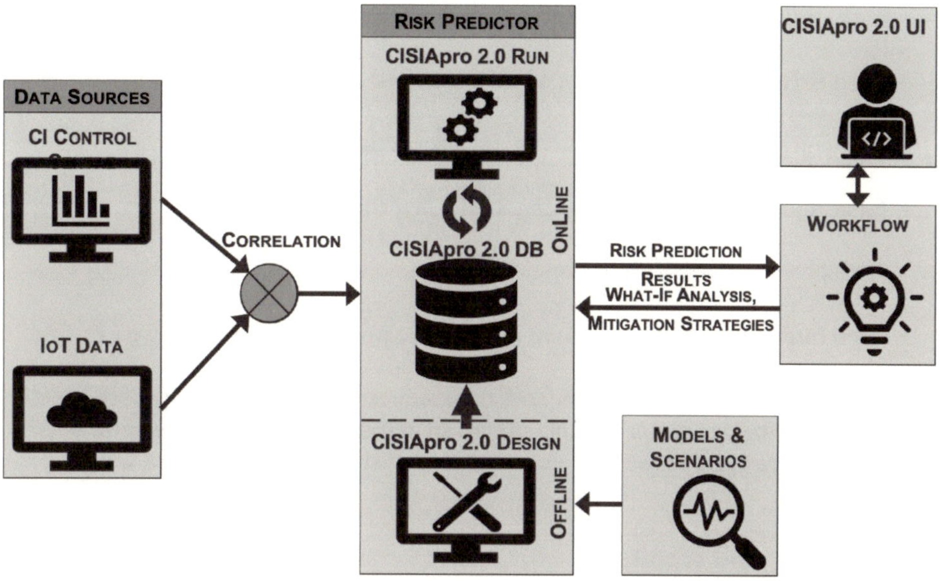

Figure 5.
CISIApro 2.0 architecture.

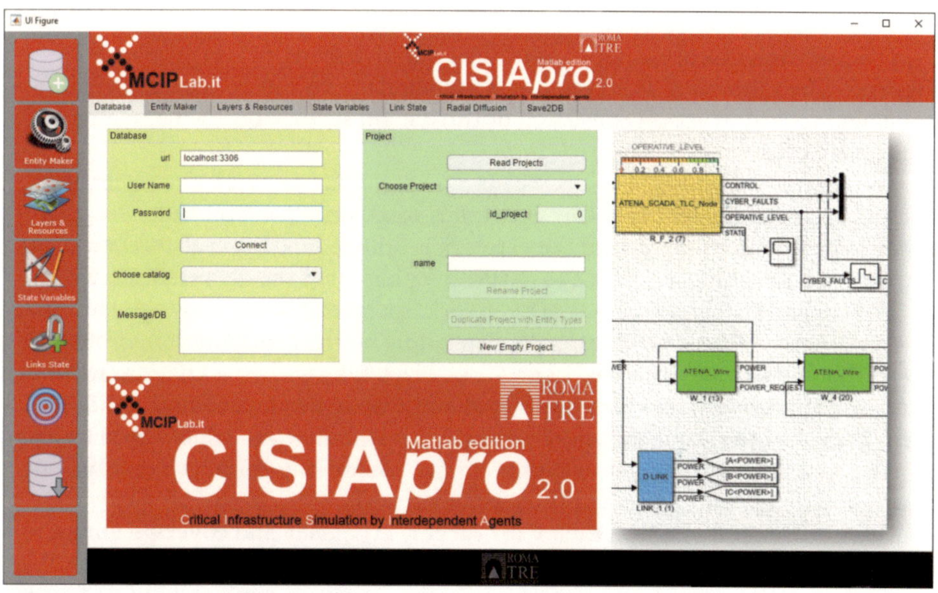

Figure 6.
CISIApro 2.0 Graphical User Interface.

plays the important role of Hybrid Risk Evaluation Tool. Hybrid because it is able to get information of different natures (sensor and data acquisition and complex event processing systems) and translating them in operational levels of resources, faults or services for the entities introduced in the critical infrastructure model.

With the proposed architecture, through CISIApro 2.0 modelling software, it is possible to dynamically change the interdependencies model and plugin other modules in order to have a pseudo-real-time scalable and flexible system, which can be changed at any time. The DB stores the information needed for the representation of several Critical Infrastructures, such as:

- Each entity is a specific instance of an entity type;
- Each entity has a status made of variables with values;
- Each entity has ports for exchanging resources;
- Each resource is associated with a MHR layer/net;
- Each layer has proper interdependencies;
- Each interconnection is made of a couple of ports, associated to two entities.

It should be noted that CISIApro 2.0 has introduced efficient ways to model, execute and debug simulations and cascading effects. In particular, an intuitive Graphical User Interface, **Figure 6**, is provided to create entities and connect them in easy way.

6. Case study and results

The proposed scenario consists of three major components: the telecommunication network, the hospital ward and the smart factory. For industrial automation and possible remote operations, the fifth generation of telecommunication networks would be an essential improvement [12].

The telecommunication network of the reference scenario is represented in **Figure 7**. The purpose of this network is to manufacture and deliver services and it has a hierarchical structure consisting of three main sectors: backbone, metro and access networks.

The Optical Packet Backbone (OPB) is a multi-service network that exchanges voice, data and video services. This network is based on IP/MPLS (Multi-Protocol Label Switching) technology and the network is fully redundant in all its components and resistant to failure conditions to ensure a high level of the delivered services.

The Optical Packet Metro (OPM) network is a metropolitan and regional collection and aggregation network capable, depending on the configuration, of managing traffic flows at the Ethernet, IP or MPLS level. Like OPB, the OPM network is a multi-service network in which both fixed and mobile services combine and, as such, guarantee the requirements of scalability, reliability, availability, and flexibility. The access network meets end-users in the telecommunications industry and greatly influences the features of the service offered.

There are several systems, each with varying efficiency and coverage zones, to build "the last mile", which is the part of the network that stretches from the client site to the first access node. The latest generation of access network (GPON-Gigabit Passive Optical Network) based on fiber optic infrastructure with OLT (Optical Line Terminal) and ONU (Optical Network Unit) is briefly described at the bottom left of **Figure 7**.

The distinctive aspect of this technology is the development of a network in which many recipients are reached by a single optical fiber: this enables you to prohibit the introduction of individual fiber ties between the control panel and the receiver, thus minimizing the cost of infrastructure.

In the central part of the figure, we have a broadband network. The strength of this technology, which has encouraged its growth and proliferation, lies in the fact that voice and data services use the same copper cables as the conventional telephone network. Data traffic received by the consumer is isolated by a splitter from voice traffic and processed by a Digital Subscriber Line Access Multiplexer

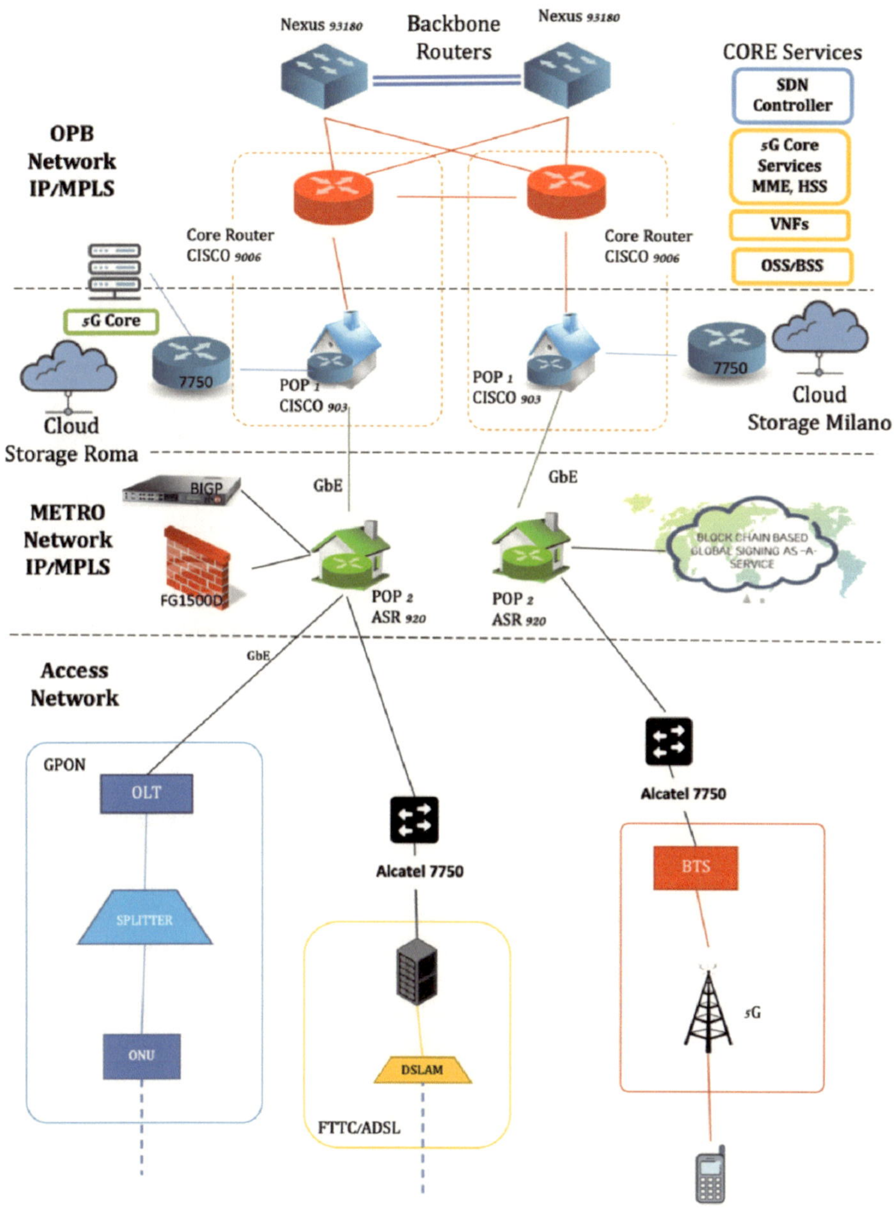

Figure 7.
The representation of the telecommunication network of the scenario.

(DSLAM) where the users' broadband lines connected to that particular central station are terminated.

On the right side of the picture, we insert the mobile network with the Base Transceiver Station (BTS) of the GSM networks that consist of antennas and transceivers responsible for the radio coverage of the territory.

The security fabric and data-center layer are achieved using a few next-generation security devices and application controllers as:

- Fortinet FortiGate (URL Filtering, Centralised Antivirus, Intrusion Detection and Protection System, E-mail filtering, Layer 4 Firewall)
- F5 BIGIP (Web Application Firewall).

Linked to the telecommunication network, we have a hospital ward represented in **Figure 8** that has been simplified to be modeled. This ward consists of a portion of the electrical grid in the yellow blocks, the water networks in blue blocks, the HVAC (Heating, Ventilation, and Air Conditioning) system in green blocks. We also add the building, made of eight rooms, where two are the operating rooms, and six are other rooms. These are the physicians' room, the staff room, the rooms used for visits, the surgery, and the waiting room, and the storage of medication and surgical supplies. These two types of rooms are modeled distinctly to underline their different relevance in the ward: while the medical and operating rooms are dedicated to patient care, must continue to provide the services requested optimally even after a failure, on the contrary, a malfunction of ordinary rooms does not drastically affect the quality of the service offered by the entire department.

The telecommunication network facilitates electrical hospital records to be processed in the clouds and relies on network-connected medical devices and systems.

Linked to the telecommunication network, a smart factor is present and is modeled in **Figure 9**. The smart factory for this scenario was modeled with reference to the radio access network architecture implemented in the factories of the future. **Figure 9** shows a completely autonomous local architecture, characterized by a pico site and an on-premises data center hub, which stores and performs data processing locally. The pico site is a small cellular base station typically covering a small area.

The 5G network is the best solution for this scenario [13, 14], which also makes it possible to incorporate the remote control of robots: according to this model, in a cloud environment, rather than in the robot itself, various functions aimed at regulating motion can be stored. It is thus assumed that the security of the networks in which the control modules work from cyber attacks is of vital importance.

The scenario contains also several services, modeled as service agents in CISIApro 2.0. Among those services, we focus our attention on the "5G Service", which is also included in **Figure 7**. 5G technology helps you to manage and control the movements of the programmable robotic arms remotely, increase human-machine interaction, capture the information processed by these intelligent systems and handle them in real-time. With regards to the hospital, the goal is to pervasively

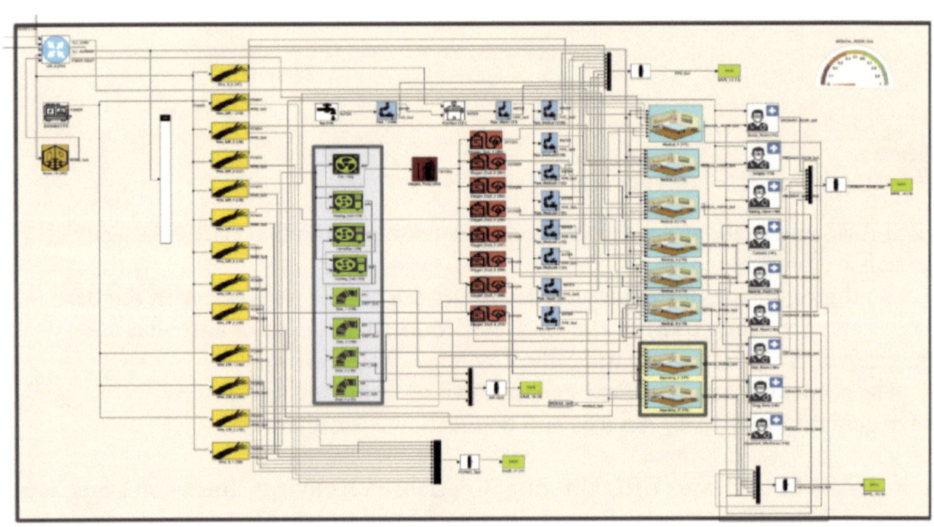

Figure 8.
The hospital in CISIApro 2.0 simulator.

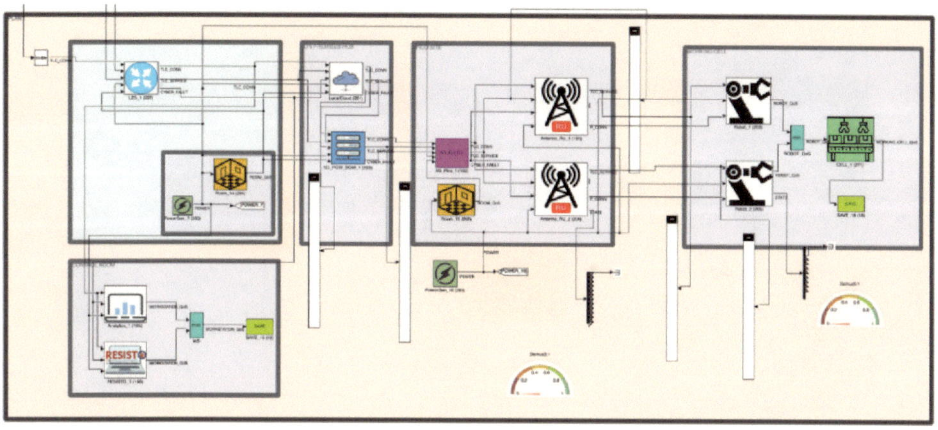

Figure 9.
The factory in CISIApro 2.0 simulator.

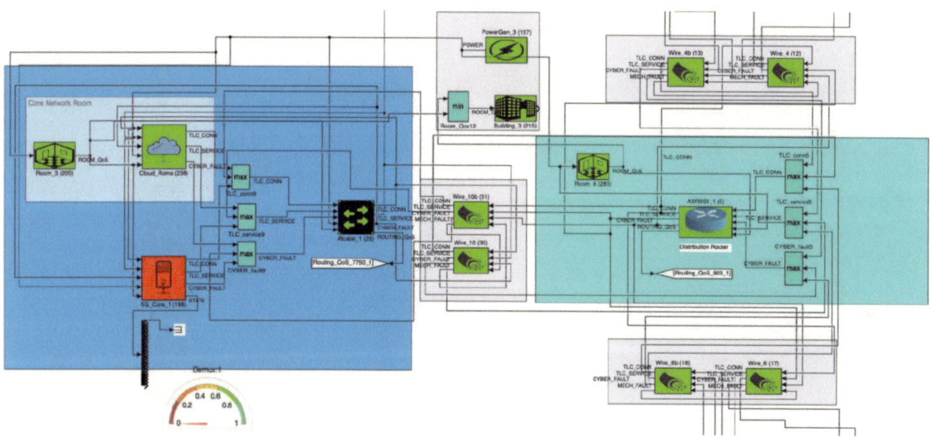

Figure 10.
The consequences on the "5G Core" component.

interconnect healthcare structures, doctors, patients, and healthcare personnel, to increase efficiency and effectiveness. In this context, the capabilities of 5G are useful for remote surgery, for remote control of the vital parameters of patients recovering from or suffering from chronic conditions and for exchanging medical data in real-time between the different technical figures.

The case study aims to examine the effects of a cyber-attack on the 5G core component, explicitly a DoS (Denial of Service). In this situation, we are not interested in how this attack was carried out, but we are more interested in the possible consequences of interconnected facilities.

The operative level of the "5G Core" agent is zero, as depicted in **Figure 10**, because it is the node that can not produce any output resource. The other entities of the telecommunications are not affected by this cyber-attack, because they don't need this service to properly work.

Different consequences affect the hospital and the smart factory. The domino effect on the smart factory is depicted in **Figure 11**. In the factory, there are four entities that need the 5G Core services to work: those entities are 5G-PGW-SGW, 5G-Pico, and the two antennas RU. Those elements are the red blocks in **Figure 11**, and they have an operative level equal to zero because they can not properly produce their outputs.

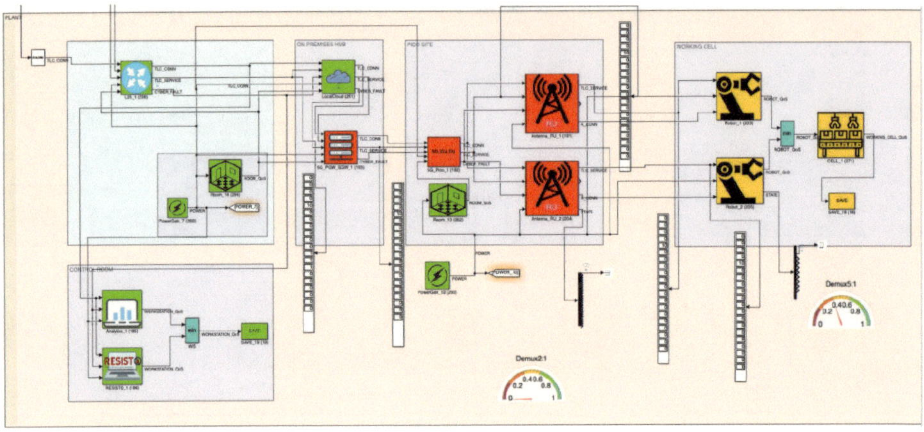

Figure 11.
The consequences on the factory section in CISIApro 2.0.

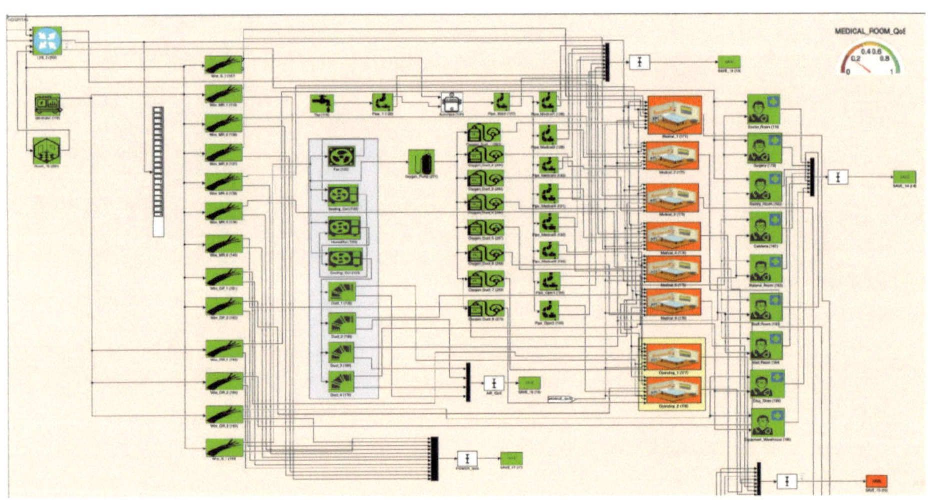

Figure 12.
The consequences on the hospital section.

Unlike the aforementioned elements, the two robots have an operative level of 0.4: although they cannot be controlled remotely or the information processed by them can be collected, however, these intelligent systems continue to operate.

In **Figure 12**, the output for the hospital is depicted. The absence of the 5G service has a more significant impact on medical rooms and operating rooms, due to the importance that hospital infrastructure has. In fact, despite following the cyber attack, it is no longer possible to carry out remote surgery, remotely monitor the vital parameters of patients and manage electronic medical records, these health rooms are still available for use and to ensure adequate care for patients.

7. Conclusions

This chapter analyses the concept of risk and resilience for critical infrastructures. The two concepts are tied together: minimizing risk means improving resilience. In critical infrastructure protection world, assessing risk is very complex due to, among the others, due to interdependency: managing risk is well-established in each infrastructure, but the risk of interconnected infrastructures is still an open problem without a single solution.

Modelling infrastructures and their interdependencies could help in managing risk and also resilience. The proposed approach is called MHR and it is implemented with CISIApro 2.0, an agent-based simulator, which assesses the consequences of events on the reference scenario. We test the proposed approach into a telecommunication scenario, with a hospital ward and a smart factory. The results demonstrate the correctness of this approach that is currently under validation within the EU H2020 RESISTO project. During the project, the system will be integrated into real test-bed provided by various telecommunication providers.

Acknowledgements

This chapter is partially supported by the European Union's Horizon 2020 research and innovation programme under Grant Agreement No. 786409 (RESISTO - RESIlience enhancement and risk control platform for communication infraSTructure Operators).

Conflict of interest

The authors declare no conflict of interest.

Author details

Chiara Foglietta*† and Stefano Panzieri†
University of Roma Tre, Rome, Italy

*Address all correspondence to: chiara.foglietta@uniroma3.it

† These authors contributed equally.

IntechOpen

© 2020 The Author(s). Licensee IntechOpen. This chapter is distributed under the terms of the Creative Commons Attribution License (http://creativecommons.org/licenses/by/3.0), which permits unrestricted use, distribution, and reproduction in any medium, provided the original work is properly cited.

References

[1] Department of Homeland Security (DHS). National Infrastructure Protection Plan: 2007/2008 Update. Technical report, 2007.

[2] European Parliament. Directive 2002/91/EC of the European Parliament and of the Council of 16 December 2002 on the energy performance of Buildings 2009.

[3] O'Rourke T, Briggs T. Critical infrastructure, interdependencies, and resilience. The Bridge. 2007;**37**:01

[4] Merriam-Webster. Resilience.

[5] Michel Bruneau and Andrei Reinhorn. Overview of the resilience concept. In *Proceedings of the 8th US national conference on earthquake engineering*, volume 2040, pages 18–22, 2006.

[6] DHS Risk Steering Committee et al. Dhs risk lexicon. *Department of Homeland Security Tech. Rep*, 2008.

[7] Kasthurirangan Gopalakrishnan and Srinivas Peeta. *Sustainable and resilient critical infrastructure systems: simulation, modeling, and intelligent engineering*. Springer, 2010.

[8] Steven M Rinaldi, James P Peerenboom, and Terrence K Kelly. Identifying, understanding, and analyzing critical infrastructure interdependencies. IEEE Control Systems Magazine, 21(6):11–25, 2001.

[9] Haimes YY, Jiang P. Leontief-based model of risk in complex interconnected infrastructures. Journal of Infrastructure Systems. 2001;7(1):1-12

[10] Giusj Digioia, Chiara Foglietta, Stefano Panzieri, and Alessandro Falleni. Mixed holistic reductionistic approach for impact assessment of cyber attacks. In *2012 European Intelligence and Security Informatics Conference*, pages 123–130. IEEE, 2012.

[11] Chiara Foglietta, Cosimo Palazzo, Riccardo Santini, and Stefano Panzieri. Assessing cyber risk using the cisiapro simulator. In *International Conference on Critical Infrastructure Protection*, pages 315–331. Springer, 2015.

[12] Mansoor Shafi, Andreas F. Molisch, Peter J. Smith, Thomas Haustein, Peiying Zhu, Prasan De Silva, Fredrik Tufvesson, Anass Benjebbour, and Gerhard Wunder. 5G: A tutorial overview of standards, trials, challenges, deployment, and practice. IEEE Journal on Selected Areas in Communications, 35(6):1201–1221, jun 2017.

[13] Sriganesh K Rao and Ramjee Prasad. Impact of 5g technologies on industry 4.0. *Wireless personal communications*, 100(1):145–159, 2018.

[14] Massimo Condoluci, Maria A Lema, Toktam Mahmoodi, and Mischa Dohler. 5g iot industry verticals and network requirements. In *Powering the Internet of Things With 5G Networks*, pages 148–175. IGI Global, 2018.

Chapter 2

Risk Analysis in Early Phase of Complex Infrastructure Projects

Esra Tepeli

Abstract

Infrastructure construction projects are complex with a very long life-cycle, a complex organizational plan, a complex resource management, technical complexities, contractual complexities and macro-environmental factors. The complexity of an infrastructure project leads to the existence of interdependent risks, which are hard to anticipate and control. As the investment is major for these types of projects, the risks and opportunities are critical to the project success or failure, the risk factors need to be identified and analyzed before any decision-making process. While upfront planning is important, not all events and scenarios can be foreseen as the project can take several years to complete and may involve many companies and stakeholders. In this planning stage of the project, a robust risk analysis method is indispensable for identifying and analyzing the major risk and opportunity factors. In this paper, a formalized multi-criteria decision-making process is developed based on a strategic risk analysis in a complex environment: (1) in a very early stage and at a strategic level, (2) before the contracting phase in order to develop a risk allocation plan and negotiate it with the project owner.

Keywords: risk management, complex projects, infrastructure projects, environmental risk analysis, risk breakdown structure, multi-criteria analysis, decision-making process

1. Introduction - risk management context for complex infrastructure projects

Risk is defined, according to ISO 31000, as the effect of uncertainty on the objectives to be achieved [1]. The last decades have been marked by notable developments in terms of infrastructure construction projects but also by unfulfilled objectives which challenge the construction industry. Strong gaps are identified in terms of organization and general management at the project level, in particular relating to the interfaces between the project actors whose specific objectives may be different or even contradict. This results in a persistent difficulty for controlling risks with the increase of the number of stakeholders. These difficulties are further heightened for complex and strategic projects. A complex and strategic project is a project that requires during its life cycle, an organization and a specific approach to manage the project, risk and opportunities [2]. Whether a project is classified as complex and strategic depends on several criteria. These criteria may relate to the organization or company which manages the project such as the level of fit with the general strategy, the main objectives of the organization, its culture and financial

 IntechOpen

state. Other criteria relate to the nature of the project such as the commercial environment, the financial plan, the brand image, the organizational plan, and the technical features. External criteria are the environmental factors such as politic factors, legal factors, social factors, international aspects when the project is abroad. External factors occur outside the organization but can lead to internal changes and are, for the most part, beyond the control of the organization [3].

Infrastructure construction projects belong to this type of strategic and complex projects as they focus on the development and maintenance of services, facilities, and systems. Infrastructure construction projects include bridges, power & energy infrastructures, roads and railroads, airports, water infrastructures and dams, and waste management plants. Infrastructure projects have a long life-cycle including the maintenance-exploitation phase. Moreover, infrastructure projects must manage complex organizational aspects, complex resource management, and complex technical and financial aspects. Such projects can be also affected easily by the environmental factors, for instance the macro-economic conditions or the politic factors of the country. These types of projects are major investment projects and can be funded by private companies, publicly, or combined as a public-private partnership (a collaboration of government entities and private sector companies). Because of all these aspects, the risk and opportunities to the project are critical and need to be identified and analyzed before any decision-making process takes place. The ITA/AITES report highlights how risk management is important in the early phase of complex tunneling projects. The report recommends a set of good practices with include the shared analysis of the risk of both the client and the potential contractors [4].

In addition, the contractual framework of infrastructure projects can be very complex; it leads to redefining the role of the project actors, their responsibilities and missions. Risk management is essential in order to identify and assess the risk and opportunity events throughout the project life cycle. Especially for the private contractor, identification of the risk and opportunity events is crucial in the early phase of the project. In this phase, the candidate contractor needs to make a strategic decision for making an offer to the tender for the project or to pull out. This decision will lead to the initial risk assessment, then offer submission with a detailed risk analysis to be able to negotiate the contract terms with the client, and to define the risk allocation plan when contract awarded [5, 6].

The risk analysis in the early phase enriches the decision-making process. The risk analysis provides rational arguments which help to avoid or mitigate the probability or impacts of negative risk events and to increase the probability and impacts of positive events which are called opportunities [7]. However, literature review shows a gap in terms of risk identification and assessment methods concerning the early phase of a complex infrastructure project [2]. The project risk identification and assessment methods in the literature consider the risk factors in a static way. Therefore, these methods have some limitations in term of adaptability and even applicability to the early phase. The difficulty is that, in the early phase of a complex project, the identification of risk events can be limited because of a scarce level of information about the project and uncertainties. With the project progresses, more information becomes available, and more precise risk identification and assessment can be performed. For this reason, developing a formalized risk management method in the early phase is necessary to identify and analyze the major risks and opportunities of an infrastructure project and to make a strategic decision for the project's future.

Therefore, the purpose of this paper is to propose a strategic and environmental risk analysis process which is applicable to the early stage and at the strategic level of an infrastructure project. In the process, the environmental risk and opportunity factors are analyzed using a formalized multi-criteria approach. This approach supports to take an optimal strategic decision for assigning some resources to a given (possible) project and later, after preliminary studies, to adequately consider detailed studies. For the client or project owner, the strategic decision corresponds to the validation of the project program and starting the step "call for tenders." For the contractor, the strategic decision corresponds to the decision to respond to the call for tenders or to pull out of the project. Then, various possible projects can be compared in order to choose and pursue the most beneficial ones, allocate the project risk optimally and control it as the project progresses. On the other hand, the strategic and environmental risk analysis process can be adapted to the evolving nature of the infrastructure project, refining the first identification of risk factors performed in the early phase of the project [8]. In the method proposed, special attention will be paid to the point of view of a private contractor with the option of adapting for multiple stakeholders if necessary.

In this perspective, Section 2 of the book chapter provides the modeling of the strategic and environmental risk analysis process in early phase of complex infrastructure projects. In the development of the process, we emphasized on a hybrid approach for the identification and analysis of risk factors which combines literature analysis, case studies of complex infrastructure projects and the Delphi technique. In Section 3, the qualitative risk assessment method and decision-making process will be explained following the principles presented in Section 2.

2. Strategic and environmental risk analysis of a complex infrastructure project

We call "project risks" as the effects of uncertainties on the project objectives in terms of time, cost, performance, quality and safety. The project risks must be managed and controlled optimally in order to achieve the project objectives. Project risk management consists of identifying risk events and analyzing them qualitatively and quantitatively. Risk analysis qualifies and/or quantifies the probability of occurrence of an identified risk and/or opportunity event and their possible negative and/or positive impact(s) on the project objectives. Finally, action plans can be proposed to the risk to a level where the residual risk is accepted. In the development of an effective risk management method, it is necessary to take into account the project objectives, the project's environmental factors and integrate the vision of the various project partners. The most classic objective of an infrastructure construction project is to manage and optimize costs and deadlines, to ensure quality and performance [9–11]. In the context of infrastructure projects, performance is understood over the long term, because it may include the entire period of maintenance and operation.

For analyzing the environmental factors in a complex project environment and understanding different stakeholders' perspectives, we followed a hybrid analysis methodology with:

1. a literature review about risk management in the early phase of infrastructure projects,

2. case studies of infrastructure projects for identifying the main risk and opportunity factors,

3. Delphi-technique sessions for understanding the perspectives of the main project stakeholders (project owner, principal contractor, consultant and other

contracted parties) about risk management in a complex project environment in the early phase, defining the process of the strategic and environmental risk analysis method, identifying and assessing main risk and opportunity factors in the early phase of the project.

The results of the literature review and the case studies revealed that in most cases of complex and strategic infrastructure projects, the main risk and opportunity factors are financial, economic, political-legal, organizational, managerial, strategic and technical factors, payment issues, construction design and technical risk, and inappropriate risk allocation across the project stakeholders [12–15].

Then, Delphi-technique sessions were carried out with the main stakeholders of infrastructure projects such as project owner, contractor, financial partners, and external stakeholders for defining a risk management strategy in the early phase.

Following the literature review, analysis of case studies and the Delphi-technique sessions, we developed the strategic and environmental risk analysis method with an external and internal risk analysis. The external risk analysis carries out the identification and analysis of the risk and opportunity factors related to the external environment of the project, such as political-legal, economic, social, technological, contractual, competitive, client's influence and force majeure factors [15, 16]. In parallel, the internal risk analysis identifies and analyzes the risk and opportunity factors related to the internal environment of the project facing the project stakeholders. These factors comprise the stakeholders' financial situation, technical strength/weakness, organizational dynamics, relationships with other project stakeholders, project client's influence, project competitors' influence and the interface between project stakeholders [17–19].

The life cycle of infrastructure construction projects can be very long with multiple phases such as feasibility studies, preliminary studies, technical studies and design, competitive dialog or tendering and contracting, administrative procedures, construction, maintenance and operation (**Figure 1**). The aforementioned risk analysis in the early phase of the project which includes strategic studies and the project's feasibility is essential for managing the risk across the whole project's life [20–21]. In these phases, project managers do not have detailed information about the project, they have only information about project scope, program and project environment. For this reason, the identification and assessment of project risks can be very challenging because of the lack of knowledge and uncertainties. Therefore, a strategic and environmental analysis can be used for identifying the main risk and opportunity factors to take a strategic decision about the project's future (GO or STOP decision) and to define a risk allocation strategy or/and preliminary risk response planning before the contracting phase (**Table 1**). The goal is to qualify the threats-opportunities and strengths-weaknesses of the project related to its environment [22–24]. Then, we detailed the risk and opportunity factors related to the external and internal environment of an infrastructure project and defined a

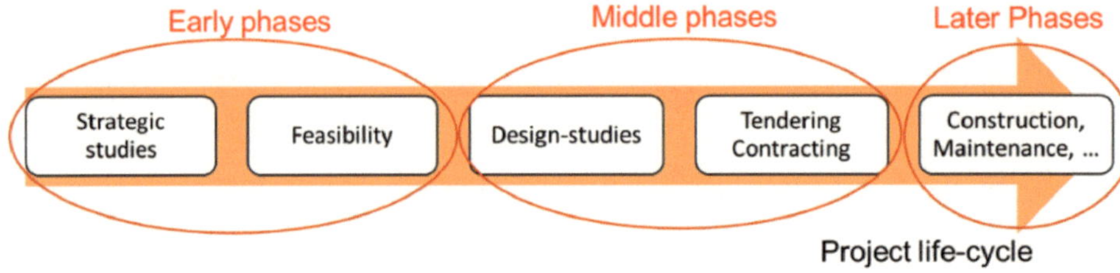

Figure 1.
Life-cycle of an infrastructure construction project.

Phase(s)	Strategic studies - feasibility
Objective	Realize the strategic analysis and environmental analysis before the decision GO/ STOP for the project, identify the risk and opportunity factors
Available information	Project scope, program, localization of the project, project life-cycle, client, commercial environment, contract information, budget, competitive environment, technical information, financial information, project life-cycle, organization, resource information, external and internal environmental factors of the project
Method /tool	Strategic and environmental analysis

Table 1.
Strategic and environmental analysis in the early phases of an infrastructure construction project.

qualitative risk assessment method to analyze the overall risk level of the project in the early phase.

Following the literature review and the Delphi-Technique sessions with the project stakeholders, in the first step, a set of risk factors is defined for both the external and internal environment as part of the strategic and environmental risk analysis process (**Table 2**). The objective is to identify the risk and opportunity factors of a complex project related to the external and internal environment, to carry out a qualitative or quantitative risk assessment in the early phases, and to make a strategic decision for the project's future. Then, a risk breakdown structure is developed with factors and sub-factors, and a qualitative evaluation method is proposed for the risk and opportunity assessment.

Project environmental factors	
1. External environmental factors	**2. Internal environmental factors**
1.1. Political-legal	2.1. Project life-cycle
1.2. Contractual	2.2. Organization
1.3. Economic	2.3. Technical features
1.4. Social	2.4. Financial features
1.5. Client influence	
1.6. Competitive environment	
1.7. Technology	
1.8 Force Majeure	

Table 2.
External and internal environmental factors.

2.1 External environmental factors

In project management, it is common to analyze the factors that are closer and more directly related to management, such as time management, resource or cost management. It will be more difficult to control the more general factors from the exterior perimeter to the project. It is therefore essential to be aware of the environmental factors that can represent restrictions and favorable circumstances in order to propose accurate risk response planning for the project success. This analysis will also apply to the project risks related to adverse environmental factors. In all cases, organizations must be prepared to mitigate the negative risk. The external environment covers the factors that can influence the project from outside the organizations [25–27]. We can distinguish the macro-environment from the micro-environment. The macro-environment analysis focuses on the broad scope that will influence the project directly or indirectly, such as political, legal, macro-economic and social factors. The micro-environment analysis highlights the interactions and relationships with other project stakeholders, the influence of the stakeholders on the project, the competitive analysis, and the technological factors. The interface between the macro-environment and the micro-environment includes lobbying, conventions, and contracts which determine the effects of the global environmental factors on the project perimeter.

In the external environmental risk and opportunity analysis, a risk breakdown structure has been elaborated with the external environmental factors and sub-factors of an infrastructure construction project (**Table 3**).

In the external environment eight risk factors are defined: (1.1) political-legal, (1.2) contractual, (1.3) economic, (1.4) social, (1.5) client influence, (1.6) competitive environment, (1.7) technology, and (1.8) force majeure.

Then, a qualitative multi-criteria evaluation takes place for assessing the risk level of the external environmental factors. As a result, a qualitative risk matrix is obtained. Each criterion is evaluated on a qualitative 5-level Likert scale [28]: High Risk, Risk, Neutral, Opportunity, High Opportunity.

Political-legal factors determine the extent to which government and government policy may impact on an organization or a specific industry as well as trade, fiscal and taxation policies, employment legislation, consumer law, trade regulation, health and safety regulations, unexpected legislation and international rules.

Contractual factors consider complexities and uncertainties which belong to the general contractual frame such as the repartition of roles and missions of stakeholders, responsibility limits and risk allocation between the stakeholders.

Economic factors influence the economy and its performance, which can give impacts on the organization and its profitability directly such as interest rates, unemployment rates, material costs and foreign exchange rates.

Social factors focus on the social environment and help an organization to understand its clients' needs and requirements. Social factors can include changing education levels, cultural trends, attitude changes and changes in lifestyles, and social security factors such as sabotage against the project, mobbing, strikes, criminal activities.

The influence of the stakeholders on the project and the relationship between project stakeholders is another external environmental factor. Mainly the client or project owner's needs must be analyzed for the project success. For the client influence factor, nine sub-factors are defined: image of the client, relations with the client, communication frequency with the client, feedback from last common projects, experience of the client for complex and strategic construction projects, project management assistance of the client, project budget allowance, financial capacity, and organizational change management-acceptance for value propositions.

The competitive factor is very challenging in the early phase for analyzing the strengths and weakness of the competitors. For the competitive environment, five sub-factors are defined: the number of competitors, competitor's size, technical capacity, financial capacity, and partners.

Technological factors indicate the rate of technological innovation and development that could affect a market or industry such as changes in technology, automation, new methods of distribution, manufacturing, logistics, research, and development. For the technology factor, three sub-factors are defined: technical difficulties, special products or innovations requested for the project and material price fluctuations.

Force majeure factor refers to an event or effect that can be neither anticipated nor controlled. There are dozens of circumstances or events that can be classed as examples of force majeure: earthquakes, hurricanes, explosions, floods, energy blackouts, epidemic diseases and war.

1. External environmental factors	Sub-factors	Qualitative evaluation				
		HIGH RISK	RISK	NEUTRAL	OPPORTUNITY	HIGH OPPORTUNITY
1.1. Political-legal						
1.2. Contractual						
1.3. Economic						
1.4. Social						
1.5. Client influence	1.5.1. Image of the client					
	1.5.2. Relations with the client					
	1.5.3. Communication frequency					
	1.5.4. Feed-back					
	1.5.5. Experience of the client					
	1.5.6. PM assistance					
	1.5.7. Project budget					
	1.5.8. Financial capacity of the client					
	1.5.9. Change management ability					
1.6. Competitive environment	1.6.1. Number of the competitors					
	1.6.2. Competitor's size					
	1.6.3. Technical capacity					
	1.6.4. Financial capacity					
	1.6.5. Partners					
1.7. Technology	1.7.1. Technical difficulties					
	1.7.2. Special products					
	1.7.3. Material price fluctuation					
1.8. Force majeure						

Table 3.
Factors and sub-factors of the external environment of a complex project.

2.2 Internal environmental factors

The internal environment covers the risk and opportunity factors that can influence the project from the inside of the organization or company. These factors comprise inter alia the features and complexities related to long project life-cycle, project management issues associated with a long life-cycle, and organizational structure. The organizational structure issues include resources, competences, communication and decision-making flows, corporate missions, corporate culture, technical features and financial properties of the project [17–19].

In the internal environmental risk and opportunity analysis, a risk breakdown structure has been elaborated with the internal environmental factors and sub-factors of an infrastructure project (**Table 4**).

In the internal environmental analysis four factors are defined: (2.1) project life-cycle, (2.2) organization, (2.3) technical aspects, and (2.4) financial aspects.

Then a qualitative multi-criteria evaluation takes place for assessing the risk level of the internal environmental factors. As a result, a qualitative risk matrix is obtained. Each sub-factor is evaluated on a qualitative 5-level Likert scale alike as in the external environmental analysis: High Risk, Risk, Neutral, Opportunity, High Opportunity.

The project life-cycle can be long for an infrastructure construction project with several phases, tasks, and milestones. For the project life-cycle factor, six

2. Internal environmental factors	Sub-factors	Qualitative evaluation				
		HIGH RISK	RISK	MEDIUM	OPPORTUNITY	HIGH OPPORTUNITY
2.1. Project life-cycle	2.1.1. Strategic studies					
	2.1.2. Design-Technical studies					
	2.1.3. Call for tenders-Contracting					
	2.1.4. Construction					
	2.1.5. Maintenance-Exploitation					
	2.1.6. Demolishing-Removal					
2.2. Organization	2.2.1. Project Management Office					
	2.2.2. Engineering Department					
	2.2.3. Construction Department					
	2.2.4. Financial Department					
	2.2.5. Legal Department					
	2.2.6. Architecture Office					
	2.2.7. Sub-contractors					
	2.2.8. Consultants					
	2.2.9. Maintainers					
	2.2.10. Suppliers					
2.3. Technical aspects	2.3.1. Technical complexity					
	2.3.2. Mastery of constructive technique					
	2.3.3. Innovation proposition					
	2.3.4. Resource availability					
	2.3.5. Quality management					
	2.3.6. Safety management					
2.4. Financial aspects	2.4.1. Financial resource					
	2.4.2. Project cost estimation					
	2.4.3. Profitability forecast					
	2.4.4. Reserves					

Table 4.
Factors and sub-factors of the internal environment of a complex project.

sub-factors are defined: strategic studies, design-technical-price studies, call for tenders-contracting, construction, maintenance-exploitation, and demolishing-removal. The objective is to evaluate the risk and opportunity factors related to the project planning and time management, the cost management for the whole project life-cycle, the complexity of tasks, and the knowledge and/or available information about the project features.

The structural organization is composed of various stakeholders with multiple organizational structures, services, and partners. There are risk and opportunity factors related to stakeholder's availability, competence, degree of experience, collaboration skills, communication skills, coordination, managerial skills and management of project resources such as resource availability, resource acquisition and transportation, resource planning and optimization.

For the organization factor, ten sub-factors are defined: Project Management Office (PMO), engineering department, construction department, financial department, legal department, architecture office, sub-contractors, consultants, maintainers, and suppliers.

For the technical features, six sub-factors are defined: technical complexity of the project, mastery of construction techniques, innovation proposition, resource availability, quality management, and safety management.

For the financial features, we can consider the factors related to financial resources, project estimation, profitability, managerial costs, and reserves. For the financial features factor, four sub-factors are defined: financial resource, project cost estimation, profitability forecast and, reserves.

3. Qualitative multi-criteria risk analysis and decision-making process

3.1 Qualitative multi-criteria risk analysis

In the definition of the project execution model, a stakeholder uses resources for realizing the project activities or tasks [2]. According to this definition, the main dimensions of a project are the project stakeholders or the structural organization, the project life cycle and the resources. The internal and external environmental factors can induce risk events which may have positive and negative consequences for the project stakeholders, resources and the project progression. In the end, these factors may impact the project objectives in terms of time, cost, quality, and safety [4] (**Figure 2**).

For instance, the macro-economic factors can influence the project funding or raw material costs; a politic or social factor can influence a stakeholder behavior; a legal factor can influence the project progression; the behavior of the public client can influence the relational flows between the stakeholders; positive public opinion about the project can induce opportunities for the project's realization.

Following the modeling of the risk breakdown structure of the environmental factors, the next step is to define a qualitative risk evaluation method to assess the external and internal environmental risk factors and to develop a global risk evaluation for the project. This assessment can be conducted at two levels:

1. in an early stage and at a strategic level for taking a strategic decision about the project,

2. before the contracting phase in order to develop a risk allocation plan.

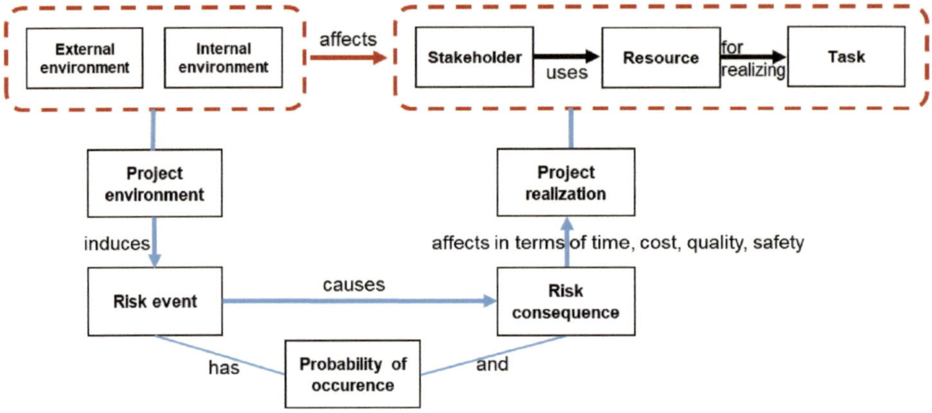

Figure 2.
Effects of environmental factors on the project realization.

The first assessment corresponds to a qualitative multi-criteria risk analysis, as part of a formalized decision-making process. The definition of the multi-criteria analysis is based on the risk breakdown structure of both sets of environmental risk factors and sub-factors. The qualitative risk assessment is realized by evaluating the environmental factors and the sub-factors in the Likert-scale from High Risk (HR) to High Opportunity (HO) as indicated in **Table 5**. In this way, we obtain a qualitative risk matrix for the external and internal environmental factors. An example of a risk matrix with the qualitative evaluations is illustrated in **Table 6**.

After codifying the Likert scale using the values in **Table 5**, the arithmetic mean "N*EnvExt*" is calculated as the risk evaluation score for the external environment of the project. Alike, the arithmetic mean "N*EnvInt*" is calculated as the risk evaluation score for the internal environment of the project. The following equations are used:

Evaluation of external environmental factors:

$$N_{EnvExt} = \frac{1}{8}\sum_{i=1}^{8} N_{cr-env-ext_i} \tag{1}$$

where $Ncr\text{-}env\text{-}ext_i$ is the evaluation score for each external environmental factor.

For the factors such as "Client influence" where there are multiple sub-factors attached, the factor's evaluation "$Ncr\text{-}env\text{-}ext_i$" is calculated using the following formula:

$$N_{cr-env-ext_i} = \frac{1}{n_{env-ext}}\sum n_{env-ext_i} \tag{2}$$

Risk/opportunity level	Score	Color code
High risk (HR)	1	Red
Risk (R)	2	Orange
Neutral (N)	3	Yellow
Opportunity (O)	4	Green
High opportunity (HO)	5	Dark Green

Table 5.
Qualitative scale for risk and opportunity levels.

1. External environmental factors	Sub-factors	Qualitative evaluation				
		HR	R	N	O	HO
1.1. Political-legal			X			
1.2. Contractual		X				
1.3. Economic			X			
1.4. Social						X
1.5. Client influence	1.5.1. Image of the client				X	
	1.5.2. Relations with the client				X	
	1.5.3. Communication frequency				X	
	1.5.4. Feed-back				X	
	1.5.5. Experience of the client				X	
	1.5.6. PM assistance			X		
	1.5.7. Project budget		X			
	1.5.8. Financial capacity of the client				X	
	1.5.9. Change management ability			X		
1.6. Competitive environment	1.6.1. Number of the competitors	X				
	1.6.2. Competitor's size		X			
	1.6.3. Technical capacity		X			
	1.6.4. Financial capacity		X			
	1.6.5. Partners		X			
1.7. Technology	1.7.1. Technical difficulties		X			
	1.7.2. Special products		X			
	1.7.3. Material price fluctuation		X			
1.8. Force majeure				X		

Table 6.
Qualitative risk matrix for environmental risk and opportunity factors.

where $n_{env\text{-}ext_i}$ is the evaluation score for each external sub-factor i, and $n_{env\text{-}ext}$ is the total number of the sub-factors attached to an external factor.

Evaluation of internal environmental factors:

$$N_{EnvInt} = \frac{1}{4}\sum_{i=1}^{4} N_{cr-env-\mathrm{int}_i} \tag{3}$$

where $N_{cr\text{-}env\text{-}int_i}$ is the evaluation score for each internal environmental factor.

For the internal factors with multiple sub-factors attached, the calculation of the evaluation score $N_{cr\text{-}env\text{-}int_i}$ for an internal environmental factor is similar as in the external environment analysis. The evaluation score of an internal environmental factor i is calculated using the following formula:

$$N_{cr-env-\mathrm{int}_i} = \frac{1}{n_{env-\mathrm{int}}}\sum n_{env-\mathrm{int}_i} \tag{4}$$

where $n_{env\text{-}int_i}$ is the evaluation score for the internal sub-factor i, and, $n_{env\text{-}int}$ is the total number of the sub-factors per internal factor.

For the risk matrix example illustrated in **Table 6**, the resulting evaluation score for the “external risk environment” of the project is:

$$N_{EnvExt} = 2.55 \tag{5}$$

If the resulting evaluation score N*EnvExt* is smaller than 3, the external environment of the project is qualified as “risky” according to the qualitative scale of Table. If the resulting evaluation score evaluation score is larger than 3, the environment is qualified as “opportune”. The evaluation score N*EnvInt* for the internal environment of the project can be calculated and assessed in a similar way.

3.2 Decision-making process

Following the analysis of the project’s external and internal environments, the risk evaluation scores help to assess if the project will be opportune or not for the company. Based on this assessment, the company can take a go/no go decision for the project. **Figure 3** shows the decision-making process. If the evaluation score of the external environmental analysis “N*EnvExt*” is larger than or equal to 3, we look at the evaluation score “N*EnvInt*” of the internal environmental analysis. If this evaluation score is also larger than or equal to 3, the project is qualified as an “opportune” project. An opportune project means that the project shows more opportunities than risk aspects. A GO or ACCEPT decision is proposed with a risk monitoring option.

If the evaluation score of the internal environmental analysis “N*EnvInt*” is smaller than 3, a brainstorming session is organized for discussing if the company can deal with the project risk when necessary risk mitigation actions are planned. If the project managers agree that there are more opportunities than risk and the negative risk impacts could be reduced with the application of action plans, they could propose a GO or ACCEPT decision.

If the evaluation score N*EnvExt* is smaller than 3, we look at the evaluation score N*EnvInt*. If this score is also smaller than 3, the project is qualified as a “risky” project, which means that there is more risk than opportunities and a STOP or REJECT decision is generally proposed. If the evaluation score is larger than or equal to 3, a brainstorming session is organized for discussing if the company can develop action plans to mitigate the negative risk impacts.

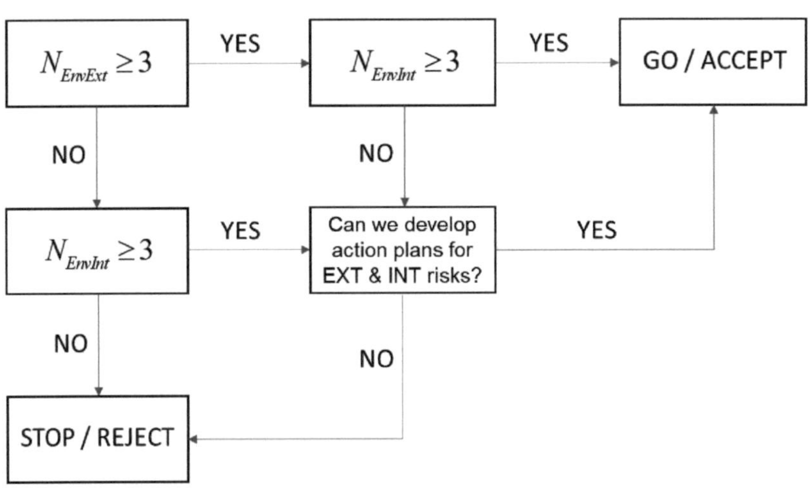

Figure 3.
Decision-making process for the environmental risk analysis.

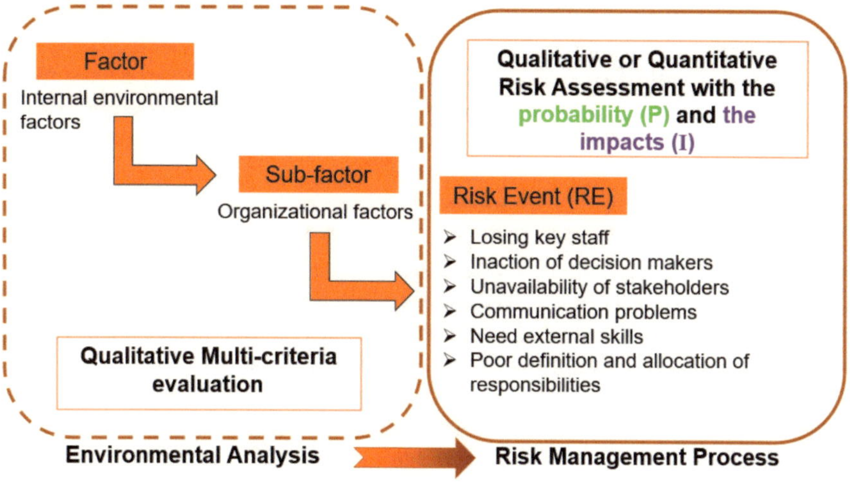

Figure 4.
Detailed risk analysis in later phases of the project based on the environmental analysis.

The strategic and environmental analysis permits to identify and assess the main opportunity and risk factors in the early phase of the project. With the help of the multi-criteria analysis, project managers can also compare multiple projects and choose the most beneficial ones for the corporate strategy.

When a GO or ACCEPT decision is taken for the project, a response planning should be developed for the risk factors deemed critical. For instance, if a potential risk is identified attached to the contractual frame of the project, this factor should be analyzed in detail. Action plans should be developed to minimize the possible legal and administrative disruptions and prepare a realistic risk allocation agreement.

In the later phases, with the project progress, more information will be available about the project. Then, the strategic and environmental risk analysis of the project evolves towards a formalized risk management process. In this approach, the risk and opportunity factors can be identified and analyzed in a more detailed structure and tracked during the project life-cycle [2, 5].

In **Figure 4** some risk events examples are illustrated, attached to the organizational factors in the internal environmental analysis, such as inaction of decision makers, unavailability of stakeholders, communication problems, poor definition and allocation of responsibilities. In this step, a formalized risk register can be developed with the risk and opportunity events, the qualitative or quantitative assessment of probability of occurrence and possible impacts in terms of cost, delay, quality and safety. Then, a risk response planning can be developed and implemented in order to mitigate the risk during the project life-cycle.

4. Conclusion

For complex projects such as infrastructure construction projects, implementing a risk management strategy is essential to achieve the project goals. It is essential to be aware of project risks related to environmental factors in order to develop the appropriate action plans. Structuring a risk management strategy that includes not only risk events but also opportunities will be beneficial for the business strategy. However, developing a robust and reliable risk management strategy can be quite difficult for complex infrastructure construction projects. Complex projects may have a long and complex life-cycle, multiple stakeholders with a complex organizational plan, and contractual complexities. For these types of projects, the identification and assessment of risks is a difficult task and may depend upon the project's characteristics and the project's environmental conditions. Since complex projects can also be of strategic importance, the early project phases play an important role in risk analysis. During this period, the project managers should analyze whether the project could be beneficial or risky to the company, carry out strategic and feasibility studies, and decide to continue or not with the project. In this step, a robust decision-making strategy should be developed for the project's future, which includes a careful analysis of the risk and possible opportunities. However, the lack of precise information about the project and a large number of uncertainties may lead to certain limitations in the reliable identification and analysis of the risk and possible opportunities during the early project phase.

This paper outlines a formalized process of strategic and project environmental risk analysis at a very early stage of a complex infrastructure construction project. Examples show how this methodology has been put into practice.

In the process, the external and internal environmental risk and opportunity factors are identified and analyzed in a formalized approach to develop an optimal strategic decision to allocate certain resources to a prospective project and later, after preliminary studies, effectively consider the project for detailed studies. Then, a qualitative multi-criteria analysis is undertaken in order to evaluate the risk and opportunity factors attached to the external and internal environment of the project and to assess the overall risk level in the early project phases. At this level, highlighting the presence of uncertainties and the lack of detailed information about the project, the risk evaluation scores cannot present a firm conclusion on the overall risk assessment. However, the methodology can provide important elements to the project management and allows risk managers to discuss in detail the risk and possible opportunities to the project. In fact, the strategic and environmental analysis should be considered as a project analysis element before any decision-making process. The environmental risk analysis may provide insight for a realistic negotiation of risk allocation with the other project stakeholders. In addition, the process may provide an accurate global vision of the project and a good understanding of the project's environmental factors. The integration proposed in the model between environmental analysis and risk management received good feed-back from project experts when applying he process in operational cases.

During later project phases, the project and risk managers can perform a more detailed risk identification and analysis; identify risk and opportunity events in a more detailed breakdown structure, assess them qualitatively and quantitatively, provide risk response planning and monitor risks during the project life-cycle. The analysis can be conducted more thoroughly when the project data permits. The formalized approach integrated into the environmental risk analysis process can provide feedback on the project, and this information could be used in the analysis of future projects.

Author details

Esra Tepeli
University of Versailles, Paris Saclay, France

*Address all correspondence to: tepeli.esra@gmail.com

IntechOpen

© 2020 The Author(s). Licensee IntechOpen. This chapter is distributed under the terms of the Creative Commons Attribution License (http://creativecommons.org/licenses/by/3.0), which permits unrestricted use, distribution, and reproduction in any medium, provided the original work is properly cited.

References

[1] ISO 31000:2018. Preview Risk Management – Guidelines. Geneva, Switzerland; 2018.

[2] Tepeli E. Formalized and Systematic Risk Management Process for Complex and Strategic Construction Projects [Ph. D. Thesis]. The University of Bordeaux, Institute of Mechanical Engineering, Department of Civil and Environmental Engineering. 2014.

[3] Michaud P, Rochet C. Maitrise d'ouvrage Stratégique de projet, SECOR, 1999.

[4] Eskesen SD, Tengborg P, Kampmann J, Veicherts TH. Guidelines for tunnelling risk management: International Tunnelling Association, Working Group No. 2. Tunnelling and Underground Space Technology. 2004. Vol. 19, Issue 3, pages 217–237.

[5] Tepeli E, Taillandier F, Breysse D. Multidimensional Modelling of Complex and Strategic Construction Projects for a More Effective Risk Management. International Journal of Construction Management. 2019. DOI: 10.1080/15623599.2019.1606493.

[6] Hwang BG, Zhao X, Gay MJS. Public-private partnership projects in Singapore: Factors, critical risks and preferred risk allocation from the perspective of contractors. International journal of project management. 2013. Volume 31, Issue 3, p. 424–33. 2013. https://doi.org/10.1016/j.ijproman.2012.08.003.

[7] Greco S, Ehrgott M, Figueir JR. Multiple Criteria Decision Analysis. State of the Art Surveys. (Book). Springer, New York, NY. 2005. https://doi.org/10.1007/978-1-4939-3094-4.

[8] Petit Y, Hobbs B. Project portfolios in dynamic environments: Sources of uncertainty and sensing mechanisms. Project Management Journal. 2010. 41, 4, (46–58).

[9] Baloi D, Price A. Modelling global risk factors affecting construction cost performance. International Journal of Project Management. 2003. Volume 21, Issue 4, pp. 261–269. https://doi.org/10.1016/S0263-7863(02)00017-0.

[10] Thamhain, H. Managing Risks in Complex Projects, Project Management Journal. 2013. 44, 2, (20–35).

[11] Thomas, A., Kalidindi, S.N., Ganesh, L. Modelling and assessment of critical risks in BOT road projects. Construction Management and Economics. 2006. 24 (4), p. 407–424. Doi: 10.1080/01446190500435275.

[12] Ke Y, Wang S, Chan A. Risk Allocation in Public-Private Partnership Infrastructure Projects: Comparative Study. Journal of Infrastructure Systems. 2010. Volume 16, Issue 4. https://doi.org/10.1061/(ASCE)IS.1943-555X.0000030.

[13] Emblemsvåg J, Kjølstad LE. Strategic risk analysis-a field version. Management Decision. 2002. Vol. 40 Issue: 9, pp.842–852. https://doi.org/10.1108/00251740210441063.

[14] Likhitruangsilp V, Tien Do S, Onishi M. A Comparative Study on the Risk Perceptions of the Public and Private Sectors in Public-Private Partnership (PPP) Transportation Projects in Vietnam. Engineering Journal. 2017. Volume 21, Issue 7. Doi: 10.4186/ej.2017.21.7.213.

[15] Yuksel I. Developing a Multi-Criteria Decision-Making Model for PESTEL Analysis. International Journal of Business and Management. 2012. Vol. 7, No. 24. ISSN 1833–3850. E-ISSN 1833–8119. Doi:10.5539/ijbm.v7n24p52.

[16] Lyons T, Skitmore M. Project risk management in the Queensland engineering construction industry: a survey. International Journal of Project Management. 2004. Volume 22, Issue 1, pages 51–61. https://doi.org/10.1016/S0263-7863(03)00005-X.

[17] Lu W, Heng Li H, Liyin Shen L, Huang T. Strengths, Weaknesses, Opportunities, and Threats Analysis of Chinese Construction Companies in the Global Market. Journal of Management in Engineering. 2009. Volume 25, Issue 4. https://doi.org/10.1061/(ASCE)0742-597X(2009)25:4(166).

[18] Beringer C, Jonas D, Gemünden HG. Establishing Project Portfolio Management: An Exploratory Analysis of the Influence of Internal Stakeholders' Interactions. Project Management Journal. 2012. 43, 6, (16–32).

[19] Voss M. Impact of customer integration on project portfolio management and its success-Developing a conceptual framework. International Journal of Project Management. 2012. https://doi.org/10.1016/j.ijproman.2012.01.017, 30, 5, (567–581).

[20] Pheng LS, Chuan QT. Environmental factors and work performance of project managers in the construction industry. International Journal of Project Management. 2006. Volume 24, Issue 1, Pages 24–37. https://doi.org/10.1016/j.ijproman.2005.06.001.

[21] Sanchez H, Robert B, Pellerin R. A project portfolio risk-opportunity identification framework. Project Management Journal. 2008. 39, 3, (97–109).

[22] Hofman M, Grela G. Taxonomy of the project portfolio risks - an empirical investigation. Procedia Computer Science. 2017. https://doi.org/10.1016/j.procs.2017.11.019, 121, (137–144).

[23] Hillson, DA. The Risk Breakdown Structure (RBS) as an Aid to Effective Risk Management. Proceedings of the 5th European Project Management Conference. PMI Europe. 2002, presented in Cannes France.

[24] Hamzaoui F, Taillandier F, Mehdizadeh R, Breysse D, Allal A. Evolutive Risk Breakdown Structure for managing construction project risks: Application to a railway project in Algeria. European Journal of Environmental and Civil Engineering. 2014. 19–2, p. 238–262.

[25] Nitank R, Trivedi MK. Pestle Technique. A Tool to Identify External Risks in Construction Projects. International Research Journal of Engineering and Technology (IRJET). 2016. Volume 3, Issue 1. E-ISSN: 2395-0056.

[26] Kim HJ, Kim BK. An entrepreneurial paradox: the moderating effect of the external environment, Asian Journal of Technology Innovation.2016. 24, 2, (222).

[27] Gupta, A. Environment & PEST Analysis: An Approach to External Business Environment. International Journal of Modern Social Sciences. 2013. 2(1): 34–43. ISSN: 2169–9917.

[28] Joshi A, Kale S, Chandel S, Pal DK. Likert Scale: Explored and Explained. British Journal of Applied Science and Technology. 2015. 7(4):396–403. DOI: 10.9734/BJAST/2015/14975.

Chapter 3

Resilience of Critical Infrastructures: A Risk Assessment Methodology for Energy Corridors

Andrea Carpignano, Daniele Grosso, Raffaella Gerboni and Andrea Bologna

Abstract

The need for scientific methodologies to assess quantitatively the resilience of critical infrastructures against natural hazards (like earthquakes, floods, storms, landslides and wildfires) during the last decade has become a relevant aspect for several countries and for the European Union. In fact, this quantification could allow setting and implementing effective measures to prevent or mitigate the negative socio-economic effects that a possible disruption of these infrastructures, caused by extreme natural events, could cause. This paper focuses, in particular, on energy corridors and proposes a new approach for evaluating their resilience, based on the definition of a criticality index able to estimate the economic damage associated to all the hazards by taking into account the spatial dimension of the infrastructure and by combining different interdependent parameters that could affect the criticality level. The procedure was tested by means of an application to a simplified case study. The obtained results highlighted the main advantages of the defined method, especially in ranking the critical sections of the infrastructure and prioritising the investments for reinforcing and protecting it or in identifying the further tests to be performed, especially in the case of a reassessment of the acceptable risk limit.

Keywords: critical infrastructures, risk acceptability, natural event, resilience, criticality index, energy corridors

1. Introduction

The reduction in the vulnerability to all the possible hazards (in many cases unpredictable) that could damage Critical Infrastructures (CIs) by improving the level of their protection and by increasing their resilience is one of the main goals of the European Union. The objective is to limit as much as possible the probability of widespread negative effects on EU's citizens and economy by ensuring services even in the case of significant disruptive events, coherently with the objectives of the Stockholm Programme [1] and of the EU Internal Security Strategy [2].

The United Nations International Strategy for Disaster Reduction (UNISDR) defined the resilience as "the ability of a system, community or society exposed to hazards to resist, absorb, accommodate to and recover from the effects of a hazard in a timely and efficient manner, including through the preservation and restoration of its essential basic structures and functions" [3]. This general statement applies also to the CIs.

According to the definition firstly given by the European Community in the 2004 Communication on "Critical Infrastructure Protection in the fight against terrorism" [4], the Critical Infrastructures are crucial systems, facilities, networks or assets which disruption would lead to relevant impacts on the socio-economic condition and development of a Member State (MS). For enhancing their protection not only against terrorism, but also against all the other hazards (thus including natural disasters), the European Programme for Critical Infrastructure Protections (EPCIP) was set [5, 6]. The aim of this programme was to define a general framework based on several principles including subsidiarity, sector-by-sector approach, complementarity, confidentiality, proportionality and stakeholder cooperation. It focused on the identification of the European Critical Infrastructures (ECI) defined as CIs located in EU's MS which disruption would significantly affect at least two MS [5]. It also addressed their possible interdependencies, the assessment of their risk by means of common approaches, the measures that could be set to improve their protection, the impacts that hazards and accidents external to EU's borders could have on the EU, the contingency plans to reduce or mitigate the negative effects of CI disruptions [5].

One of the most relevant documents for the implementation of the ECIP is the 2008 Directive on "the identification and designation of European critical infrastructures and the assessment of the need to improve their protection" [7]. It represents the first approach to identify ECI and to evaluate the need for increasing their protection level, and it refers to only two specific sectors (energy and transport), pointing out the necessity of future reviews meant to include other sectors, like the information and communication technology (ICT) one. It also requires owners/operators of the identified ECI to produce Operator Security Plans (OSP), which define the options existing or being implemented for the ECI protection.

In 2013, a revision of the EPCIP was introduced [8], aiming at organising the implementation of the activities along three work streams (prevention, preparedness and response), at deepening the analysis of the interdependencies (both cross-sector and cross-border) and at taking into account critical ICT infrastructures and their relationship with other CIs (especially electricity generation and transmission infrastructures).

In 2017, an evaluation aiming at assessing the implementation of the 2008 Directive and focusing on its relevance, coherence, effectiveness, efficiency, EU added value and sustainability has been launched by the European Commission. The assessment process ended in 2019. It puts into evidence the need of revising the Directive, including further sectors besides the energy and transport ones and taking into account the interdependencies among sectors. Furthermore, it highlights the relevance that new threats – including those related to the artificial intelligence, the introduction of advanced ICT solutions that can create new vulnerabilities and the involvement of third countries in the ownership and operation of CIs – can assume [9, 10].

In order to effectively enhance the protection of CIs, quantitative methodologies, able to evaluate their resilience and to assess, in a holistic way, the different dimensions involved are needed. In particular, the approaches proposed in the scientific literature focus on some key aspects related to the concept of infrastructure resilience, namely: ad hoc risk assessment methodologies for quantifying the resilience of CIs, interlinks and interdependencies among CIs, analysis of the infrastructure vulnerability with respect to different kind of threats. Some of these approaches also try to assess the multi-dimensional (energy, social, environmental and economic) impacts due to disruptive events involving CIs.

With respect to these aspects, different reviews of the proposed studies are available in literature, as those carried out by Ouyang [11], Griot [12], Wang et al. [13] and Liu et al. [14].

Considering the quantitative methodologies for evaluating the resilience of CIs, two studies prepared by the JRC can be firstly mentioned. In particular, Galbusera et al. [15] proposed a feasibility study for the application of stress tests (like those adopted in the nuclear and economic sectors) to the evaluation of CI resilience against several hazards. Giannopoulos et al. [16] carried out an analysis of the state of the art related to the risk assessment methodologies that could be useful for the protection of CIs. A general approach to risk analysis and management of system-of systems can be found in the studies performed by Haimes et al. [17] and by Ariel Pinto et al. [18]. Eusgeld et al. [19] analysed instead the alternative modelling options (integrated and coupled models) for system-of-systems and proposed a specific High Level Architecture (HLA) for modelling Supervisory Control and Data Acquisition (SCADA) and "System under Control" (SuC, like gas supply system or power supply system). Labaka et al. [20, 21] suggested a holistic framework (based on the identification of resilience policies, on their influence and on the methodology for their implementation) aiming at increasing the resilience of CIs by identifying their resilience level, their weaknesses and the possible improvements to be implemented. Mao et al. [22] highlighted that different measures aiming at increasing the resilience of CIs can be coherent or conflicting among each other, due to a missing systemic approach. Consequently, they proposed a framework based on a quality function deployment (QFD) that takes into account the correlations between resilience improvement actions at different stages of the CIs lifecycle. Nan et al. [23] proposed a method for resilience estimation, which combines a hybrid multi-layer model (for capturing the interaction between different subsystems) and an integrated metric (for the quantification of the resilience, considering the different resilience capabilities). Ouyang et al. [24] focused on the CIs protection, starting from the actions that can be adopted to protect weak system components before a disruptive event happens and comparing the robustness-based approach (mainly related to the remaining functionality level of the system after the event and before the restoration) and the resilience-based approach (which includes the possible restoration path and the related rapidity).

The opportunity to model infrastructure networks as interconnected system-of-systems in order to properly describe the cascade effects due to their strong interdependencies has been underlined by several authors. Theocharidou et al. [25] suggested a new methodology – called CRitical Infrastructures & Systems Risk and Resilience Assessment Methodology (CRISRRAM) – developed in an all-hazard perspective and based on a system-of-systems approach (a definition of system-of-systems can be found in [26]), which introduces three layers (society, asset and system) and evaluates the direct or indirect effects on economy, environment and citizens caused by the hazards considered in each scenario. Another approach based on the system-of-systems concept, a Monte Carlo simulation and a Hierarchical Graph representation of the interdependent CIs is the one described by Ferrario et al. [27], which was applied to two case studies – concerning respectively small electric and gas grids (plus a SCADA system) and a large electrical distribution network – for the evaluation of their robustness. Kröger et al. [28] and Zio [29, 30] furtherly suggested an approach – helpful in CI protection – based on the risk and vulnerability concepts and able to allow the identification of possible vulnerabilities (both evident and hidden), thus avoiding the failures that could originate when the CIs are subject to hazards of multiple nature. Johansson et al. also focused on the opportunity to use vulnerability analyses to complete reliability studies of CIs [31] and demonstrated it by applying a Monte Carlo approach for reliability analyses and a vulnerability analysis to an electric power system. Moreover, Johansson et al. [32] proposed a model that could be useful in the framework of vulnerability analyses of interdependent infrastructures that are described by both a network model (based on the graph theory) and a functional model. Stergiopoulos et al. [33] explored the interdependencies among CIs that cause cascading effects in the case of failure. For this purpose, the authors started from the dependency risk methodology proposed by Kotzanikolaou et al. [34, 35] and introduced graph centrality metrics in order to identify the nodes that mainly affect the risk paths and that can thus be controlled in order to improve risk mitigation. Furthermore, Stergiopoulos et al. [36] extended the studies performed by Kotzanikolaou et al. [34, 35, 37] by considering the time evolution of each dependency (using fuzzy models) and the concurrent common-cause cascading failures, developing a supporting tool for decision making (named CIDA, i.e. Critical Infrastructure Dependency Analysis). This tool can be useful in assessing the CI's resilience under different scenarios and the effectiveness of possible mitigation actions. Fu et al. [38] also focused on the opportunity of treating infrastructure networks as interdependent system-of-systems, while Utne et al. [39] proposed a methodological approach to model the interdependencies among CIs built starting from the use of relatively simple cascade diagrams. Furthermore, the JRC developed the Geospatial Risk and Resilience Assessment Platform (GRRASP), a graphical tool for analysing network systems that can be adopted to identify the critical elements of the network and to evaluate the cascading effects of CI disruptions, taking into account cross-sectoral and cross-border interdependencies [40].

Finally, with reference to the impact analysis of different threats on CIs, specific models have been developed in order to assess the physical security and the resilience of CIs themselves against single kinds of hazards. In particular, Khalil et al. [41] focused on the modelling of physical security of CIs under attack scenarios by using a Monte Carlo-based probabilistic dynamic approach. Urlainis et al. [42] implemented instead a supporting tool for decision making suitable to evaluate the risk related to oil & gas critical infrastructures after the occurrence of a seismic event. This tool adopts fault-trees, decision trees and fragility curves and allows the identification of the most critical sections of the analysed system based on the damage state of its components. Shakou et al. [43] proposed a framework for increasing the resilience of CIs with respect to climate change phenomena, based on different timescales and promoting flexibility, modularisation and diversification.

In comparison with the mentioned studies available in the scientific literature, the new methodological approach proposed in this paper mainly focuses on single large infrastructures (like energy corridors for oil and gas supply) and aims at taking into account their geographical dimension, allowing analyses characterised by a high spatial granularity. Furthermore, the proposed procedure is able to consider the most relevant interdependencies among the parameters that could impact on the criticality of an infrastructure with a simple mathematical formulation. Therefore, this work aims at being a supporting tool not only for infrastructures management companies and for the civil protection but also for public administrations.

The paper considers the energy CIs: according to the 2008 EU Directive, this category includes facilities and infrastructures for power generation and transmission, for oil and gas production, treatment, storage and transmission and LNG terminals [7]. In particular, it focuses on the energy corridors (oil and gas pipelines, power lines).

Its goal is to define a methodology for the evaluation of a criticality index, related to the failure of an energy infrastructure due to extreme natural hazards like earthquakes, floods, storms, landslides and wildfires. This criticality index is useful to assess the criticality level of each section of the infrastructure itself (taking into account its spatial dimension) with respect to the socio-economic damage (measured in economic unit) caused by the failure. Furthermore, the possibility to estimate the distance from the criticality status even in case of non-critical scenarios and to compare the criticality condition with a risk acceptability criterion (identifying – for the most critical sections – the need for undergoing structural tests) could give a valuable support in prioritising investments and in defining suitable countermeasures and protective actions.

2. Methodology

The proposed approach starts from the concept of energy corridor. A corridor can be defined as an extensive infrastructure (like natural gas and oil pipelines and large power lines), characterised by a start point and an end point, that links production/refining facilities with distribution hubs. Energy corridors are usually strategic elements for the economy of the countries that are connected to them, and their influence spreads over a large area not limited to the geographical neighbourhood of the infrastructure. In a future world that is expected to be increasingly interconnected with large scale energy markets, the role of energy corridors could become crucial: the diversification of the sources and the possibility to ensure the functionality of the infrastructures could significantly impact on the security of energy supply and on the economic systems of several countries, especially those characterised by a high level of energy import dependency.

For these reasons, the quantitative evaluation of the resilience of the energy corridors against possible adverse events through the numerical estimation of their criticality level and the simultaneous identification of suitable criteria for risk acceptability are essential in order to identify the sections that require attention and investments for preventing potentially severe failures which could impact on the GDP (Gross Domestic Product) with losses at different scales.

According to the methodology described in the following sections, a set of parameters influencing the criticality status of the corridor and their interdependencies have been firstly defined (Section 2.1). A relationship linking these parameters has then been built to define a new Criticality Index (Section 2.2). A criterion for the risk acceptability (Section 2.3) and the application of the whole procedure to a simplified case study have been eventually discussed (Section 3).

2.1 Identification of the parameters and their interdependencies

The proposed methodology focuses on the quantitative assessment of the criticality of a single section of an energy corridor under an all-hazard perspective, i.e. with respect to all the possible extreme natural events.

For this purpose, the first step has been represented by the definition of a set of parameters that could affect the criticality level of an energy infrastructure, by their clustering into different groups and by the analysis of their interdependencies. Moreover, in order to take into account the spatial dimension of the energy corridors, the possible dependency of each parameter on the geographical position z_c (ranging between 0 and the corridor length l_c and measured in km) along the corridor itself has been explored. In fact, an infrastructure like a pipeline can typically run over long lengths and the natural environment surrounding it could significantly change along the route: consequently, certain natural hazards could be considered only for a limited set of branches and not for the overall length of the corridor. Eventually, the effects of a variation in the value of each parameter on the damage have been estimated. In particular, in this study 15 parameters and 4 groups

("Event related", "Corridor related", "Backup sources related" and "Users related") have been considered: the parameters taken into account are listed in **Table 1** and the dependency matrix is shown in **Table 2**. The interdependencies are identified assuming as increasing the value of each independent parameter and reporting the effect on the dependent parameter (decreasing or increasing when the independent parameter increases). The table reports also the effect of each parameter on damage.

Referring to Group 1, the seasonality s – that represents the variability of the considered natural event across the year – is the parameter that mainly affects the other ones. The probability p that the natural event could have an impact not only on the analysed corridor but also on other infrastructures supplying the same commodity (backup sources) is strictly related to the magnitude of the event itself and on the geographical context: it depends on the distance between the corridor (or corridor branch) and the considered backup source and on the potential damage area for the considered event, quantified through the damage distance λ. All the facilities located at a distance lower than or equal to λ are certainly involved by the event to such a degree that their functionality is lost.

In general, an increase in all the parameters related to the corridor (Group 2) causes an increase in the potential damage. It has to be highlighted that RT – which includes not only the time needed to repair the infrastructure but also the time for reaching the damaged section of the corridor and the time to get the requested spare parts – depends not only on the season but also on the temporal and spatial scale of

Group	Parameter	Description	Unit
1. Event related			
	p	Probability to involve more than a single facility	—
	λ	Damage distance (measure of the potential damage area of the event)	km
	τ	Time scale of the event (measure of its duration)	s
	s	Seasonal factor (influence of the season on the event)	—
2. Corridor related			
	l_c	Length of the corridor	km
	$c_{p,c}$	Peak capacity of the corridor	GJ/s
	RT	Repair time	s
3. Backup sources related			
	d_b	Distance between a single source and the corridor	km
	$c_{p,b}$	Peak capacity of the source	GJ/s
	$r_{m,b}$	Minimum available reserves for the single source	GJ
	α_b	Availability of the source	—
	α_{tec}	Technical availability of the source	—
4. Users related			
	i	Interruptible capacity	GJ/s
	α_i	Availability of interruptible capacity	—
	e	Energy intensity for the considered commodity	€/GJ

Table 1.
Considered parameters by group.

Parameter	Description	Dependency on the position z_c	Effects on damage		Inter-dependencies	
			↑	↓	↑ with	↓ with
p	Probability to involve more facilities	X	X		λ	d_b
λ	Damage distance		X			
τ	Event time scale		X		s	s
s	Season					
l_c	Corridor length	X	X			
$c_{p,c}$	Corridor peak capacity		X		s	s
RT	Repair time	X	X		τ, s	s
d_b	Distance source-corridor	X		X		
$c_{p,b}$	Source peak capacity			X	s	s
$r_{m,b}$	Minimum reserve of the source			X	s	s
α_b	Availability of the source	X		X	s, d_b	λ, s
α_{tec}	Technical availability			X	s	s
i	Interruptible capacity			X	s	s
α_i	Availability of i			X	s	s
e	Energy intensity		X			

Table 2.
Interdependencies and effects on damage.

the event: the greater the geographical extension of the natural event and its duration, the longer the time needed to reach the damaged section.

As it can be reasonably expected, an increase in the parameters related to the availability of backup sources causes a decrease in the damage. It can be underlined that the average distance between the backup sources provides information about the probability that a backup source could be involved in the considered extreme event: in fact, the higher the value of this parameter, the lower the probability. The availability of these sources depends not only on the seasonality, but also indirectly on the distance between the corridor and the source: in particular, it increases if the source is far from the epicentre of the event.

Considering Group 4, the parameters are related with the reference market: in case of a possible corridor failure, the market operator could decide a supply interruption for some selected users, in order to reduce the load of the considered infrastructure; the interruptible capacity could depend on the season. The energy intensity e (i.e. the amount of energy needed to produce a unit of GDP), instead, gives a measure of the importance of the commodity delivered by the considered corridor, allowing to quantify the economic damage deriving from the supply lost as a consequence of an extreme event.

It can be highlighted that the event related parameters can be evaluated on the basis of geological surveys and studies on natural hazards with respect to the specific site analysed. Among them, the probability of involving more facilities needs *ad hoc* formulations and cannot be generically expressed by means of a single mathematical relationship (as further discussed in Section 2.2). The majority of the corridor related and the backup sources related parameters are instead technical data that are usually available for the specific infrastructures considered. Only the repair time should be estimated by means of suitable databases or specific investigations (Maintainability Analyses). Eventually, referring to the users related parameters, the interruptible capacity is an information that should be known as depending on already signed contracts and agreements, while the energy intensity for the commodity carried by the corridor can be obtained from statistical sources.

Furthermore, for the proposed method, the corridor can be assumed as one-dimensional, i.e. only characterised by the running coordinate z_c. This is because only the position along the corridor, the distance between the backup sources with respect to the corridor and the distance between the epicentre of the considered natural hazard and the corridor itself are relevant for the analysis.

2.2 Definition of the criticality index

Starting from the parameters and interdependencies identified in Section 2.1, in order to define a criticality index able to quantify the criticality of a single branch/corridor, a relationship expressing the socio-economic damage D due to a certain extreme natural hazard has been defined (Eq. (1)). It expresses the damage D in the section of the branch/corridor identified by the coordinate z_c (running over the corridor length, from 0 to l_c).

$$D(s,p,z_c,\tau) = \left\{ RT(s,z_c,\tau) \cdot \left[c_{p,c}(s) - \alpha_i(s) \cdot i(s) - \sum_b \alpha_b(s,p) \cdot c_{p,b}(s) \cdot \left(\frac{T_b}{RT(s,z_c,\tau)} \right) \right] \cdot \frac{1}{e} \right\} \tag{1}$$

where:

$$\begin{cases} T_b = T_b(s,z_c,\tau) = RT(s,z_c,\tau) & RT(s,z_c,\tau) \leq \frac{r_{m,b}}{c_{p,b}} \\ T_b = T_b(s) = \frac{r_{m,b}(s)}{c_{p,b}(s)} & RT(s,z_c,\tau) > \frac{r_{m,b}}{c_{p,b}} \end{cases} \tag{2}$$

$$\alpha_b(s,p) = \alpha_{tec}(s) \cdot [1 - p(z_c)] \tag{3}$$

Eq. (1) defines the economic value of the share of the commodity carried by corridor c over the emergency time period (identified by RT) that cannot be directly delivered notwithstanding the contribution of interruptible users and the availability of backup sources. In fact, focusing on the square bracket in the equation:

- the term $c_{p,c}$ identifies the maximum amount of commodity that can be delivered per second in season s and that is lost due to the failure; as a consequence, the product between $c_{p,c}$ and RT defines the amount of energy unavailable during the repair time after the adverse event that caused the corridor failure

- the product between α_i, i and RT defines the part of this supply that can be avoided during the emergency due to the fact that some users are interruptible

- the product between α_b, $c_{p,b}$ and T_b corresponds to the amount of energy commodity that can be certainly supplied by the backup sources during the repair time.

Referring to the probability that the event could involve other facilities (in particular, the backup sources) than the considered corridor, this can be expressed by several relationships or by more complex considerations that do not allow a simple mathematical formulation according to the different classes of natural events. For example, in the case of a river flood, p is a function not only of the distance between the corridor and the facility but also of the distance between the river and the facility. Furthermore, p is equal to 0 if the considered facility is outside the boundaries of the natural hazard, regardless of the distance between the source and the corridor. A possible relationship that can be adopted for some classes of events, like earthquakes, is the one expressed in Eq. (4) where the possible involved facilities are supposed to be the backup sources b. If the distance between the backup source and the corridor d_b is lower than the damage distance λ, the facility is assumed to be certainly involved by the event. If the distance d_b is higher than λ (i.e. the facility is located outside the potential damage area) the probability that the facility is involved by the event decreases in a proportional way with the increase of d_b.

$$p(z_c) = \begin{cases} \dfrac{\lambda}{d_b(z_c)} & d_b(z_c) \geq \lambda \\ 1 & d_b(z_c) < \lambda \end{cases} \tag{4}$$

Moreover, it has to be highlighted that Eq. (1) is defined if

$$c_{p,c}(s) - \alpha_i(s) \cdot i(s) - \sum_b \alpha_b(s,p) \cdot c_{p,b}(s) \cdot \left(\frac{T_b(s)}{RT(s, z_c, \tau)} \right) > 0$$

as, from the risk analysis point of view, the damage D has to be positively defined. A negative value of D means that the corresponding corridor section is not critical: negative values of this term could be obtained, for instance, in the case that no other facilities are involved by the natural event and the loss of corridor capacity is completely supplied by backup sources.

For this reason, the proposed relationship for defining the criticality index CI as a function of the socio-economic damage is the one reported in Eq. (5):

$$CI = \begin{cases} [1 + D(s,p,z_c,\tau)] \cdot \left[1 + e^{-D(s,p,z_c,\tau)}\right] - 1 & D(s,p,z_c,\tau) \geq 0 \\ \dfrac{1}{1 - D(s,p,z_c,\tau)} & D(s,p,z_c,\tau) < 0 \end{cases} \tag{5}$$

In this case, CI does not correspond to an economic value of the damage caused by the considered event (like D), but it allows to associate a numerical value also to the corridor sections that are not strictly critical (i.e. those for which D is negative) thus measuring their "proximity" to a real potential damage and ranking them according to a criticality perspective, as the safety margins progressively reduce when a negative value of D approximates to 0.

As it can be noticed, the CI relationship is built in order to have $\lim_{D \to \infty} CI = D$ and $CI = 1$ for $D = 0$ (i.e., when the infrastructure status changes from "non-critical" to "critical").

A graphical representation of CI as a function of D can be observed in **Figure 1**.

2.3 Criteria for risk acceptability

In the scientific literature, few studies are available to identify risk acceptability criteria for the socio-economic risk, and the differences among the economic

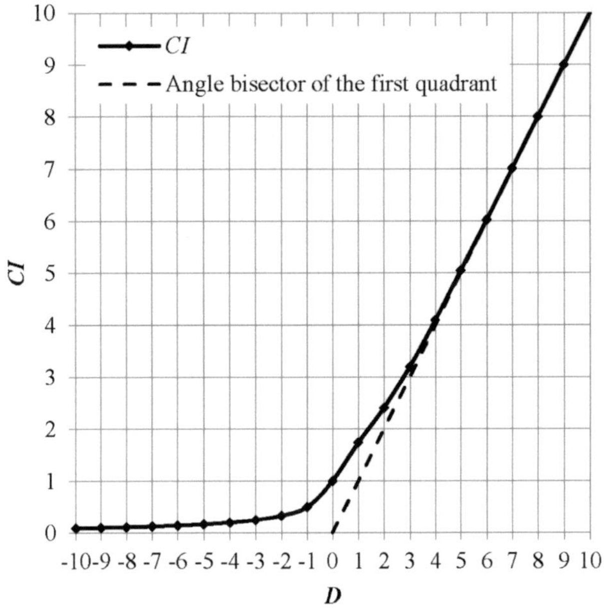

Figure 1.
Graphical representation of CI *as a function of* D.

systems do not allow to define easy procedures suitable to be applied to different contexts (like developed, developing and less developed countries).

For this reason, in the present paper a specific criterion has been proposed, based on the overall economic estimation of damages due to natural events, which takes into account both direct (i.e. to houses, infrastructures, industrial facilities, etc.) and indirect (i.e. productive losses, lack of basic services to population) damages.

According to the Munich Re insurance company statistical data, related to the global natural loss events worldwide (including geographical, meteorological, hydrological and climatological events) over the period 1980–2015 [44], the 2015 overall losses accounted for about 0.14% of the global GDP (GDP data from World Bank statistics [45]). However, during previous years significantly higher percentage values have been reached, in particular in 2011 (mostly due to the Tōhoku earthquake and tsunami in Japan), when the losses peaked at about 380 billion US dollars, and in 2005, mainly related to the hurricane Katrina in the U.S.. These two events, in particular, highlight that extreme events involving developed countries generally lead to more relevant economic effects even at a global scale.

The proposed expression for the acceptable annual economic damage related to a certain corridor is evaluated as a fraction of the annual GDP, by taking into account the contribution of the energy sector to the GDP composition, the contribution of the analysed corridor to the overall energy supply of the country/area, the weight of the economic losses due to an extreme natural event.

In particular:

- The contribution of the energy sector to the GDP is expressed by the f_{en} factor, defined as:

$$f_{en} = \frac{VA_{en}}{GDP} \tag{6}$$

where:

VA_{en}: value added of the energy sector; it has to be noticed that the GDP at market prices is the sum of the gross value added at market prices for all the productive sectors [46, 47].

- The contribution of the analysed corridor to the regional energy supply is given by the economic value of the commodity carried by the corridor c per year; the factor f_c, is defined as:

$$f_c = \frac{EV_c}{VA_{en}} \tag{7}$$

where:

EV_c: economic value of energy commodity delivered by corridor c

- The annual value of economic losses and expenditures related to the failure of the corridor c due to the natural event ne is assumed as the maximum acceptable risk, and the factor f_{ne} is defined as:

$$f_{ne} = \frac{L_{ne}}{GDP} \tag{8}$$

where:

L_{ne}: total economic losses and expenditures due to the natural event ne.

As no statistical data is available to evaluate the expenditures and economic losses for a specific natural event ne causing the failure of corridor c, the average value f_{ne}, defined at regional/country scale, is used as equivalent of the "local" ratio between the annual economic losses and expenditures associated to the failure of corridor c and the economic value EV_c of the commodity carried by c per year.

The previously described steps can be summarised into a single relationship (Eq. (9)), which allows to quantify the current economic risk in terms of monetary losses as a consequence of the adverse natural event ne:

$$R_a = f_{ne} \cdot f_{en} \cdot f_c \cdot GDP \tag{9}$$

It has to be highlighted that specific estimations of the total economic losses and expenditures L_{ne} are not commonly available as public data and should be provided by insurance companies.

Once the current risk is defined, the maximum tolerable frequency (number of events per year) for a given damage in the corridor section identified by the coordinate z_c is assessed by adopting a graphical approach which starts from the previously defined Criticality Index (i.e. the economic value of the damage caused by the service disruption due to the analysed event) (**Figure 2**).

From the obtained maximum acceptable frequency, the corresponding event intensity can be evaluated using the frequency-intensity curve, which is characteristic for each class of events (**Figure 3**).

Several studies are available in literature regarding the relationship between the frequency and the intensity (or magnitude) of natural events. For example purpose, the ones performed by Hungr et al. [48], Jakob et al. [49, 50], Riley et al. [51] (related to the debris flow landslides), Hooke [52], Zhang et al. [53] (focusing on floods), and Papadakis [54] (considering earthquakes in Greece) can be mentioned.

In general terms, the intensity is associated to specific characteristics of the considered event (like the peak ground acceleration for the earthquakes, the maximum water level for floods, the maximum wind speed for storms and the heat flux

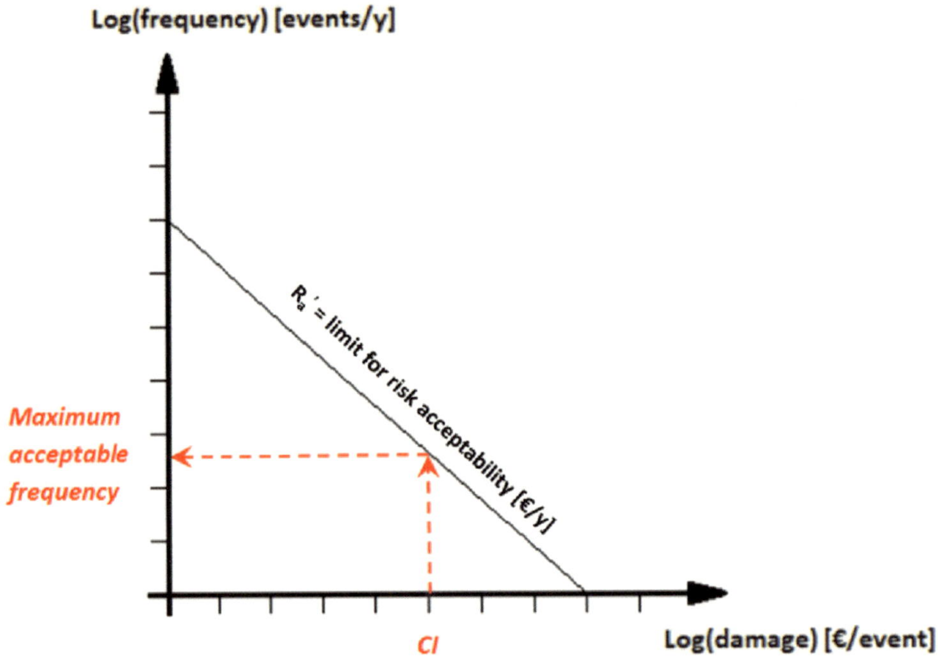

Figure 2.
Identification of the maximum tolerable frequency according to the CI *value.*

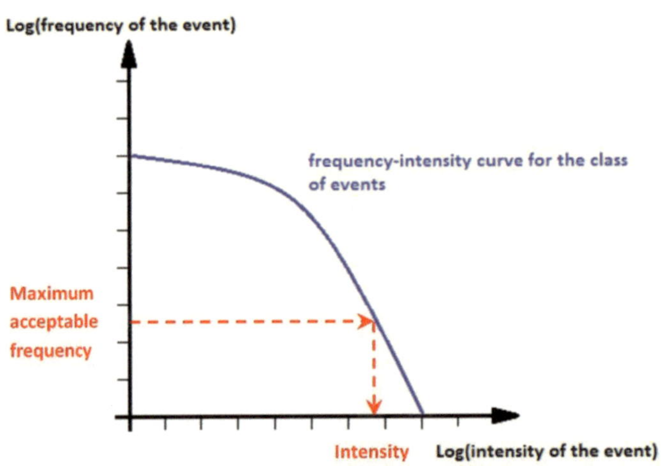

Figure 3.
Evaluation of the event intensity related to the maximum tolerable frequency according to frequency-intensity curve.

for fires) and the link between intensity and frequency is evaluated on the basis of historical data analyses.

The obtained intensity has to be compared with the design limit value for the analysed infrastructure.

It has to be further underlined that R_a represents the current overall risk related to the event *ne*. If a lower limit for risk acceptability for that event is desired, a reassessment (i.e. a reduction) has to be performed, according to Eq. (10).

$$R_a' = \alpha_{ne} \cdot R_a \tag{10}$$

where:

R'_a: reassessed limit for risk acceptability (see **Figure 2**)

α_{ne}: reassessment factor for the definition of the limit for risk acceptability related to the class of natural events *ne*; $\alpha \in [0,1]$

In this case, the same *CI* value corresponds to a lower maximum acceptable frequency, which – in turn – corresponds to a higher intensity that could exceed the design conditions of the infrastructure. In such a situation, new structural analyses have to be performed in order to verify its resilience and the possible need for mitigation actions, such as structural reinforcement, redundancy or relocation.

3. Case study and results discussion

The methodological approach described in Section 2 has been tested by applying it to a simplified case study. The main assumptions adopted can be summarised as follows:

- an ideal corridor and related surrounding environment have been taken into account;
- only two classes of extreme natural events (river floods and earthquakes) have been considered;
- three backup sources are available, able to cover the load for the entire period of unavailability of the corridor; these alternative sources are independent from the corridor itself;
- there is no interruptible capacity;
- a reassessment of the limit for risk acceptability has been assumed, with a risk reduction of one order of magnitude.

The spatial layout of the corridor and of the backup sources is shown in **Figure 4**, while their characterisation and the values of the main parameters are reported in **Table 3**.

It has to be underlined that, in this simplified case study, the values of the parameters have been chosen in order to be realistic but they are not corresponding to a real case. In particular, all the parameters have been assumed to be seasonally

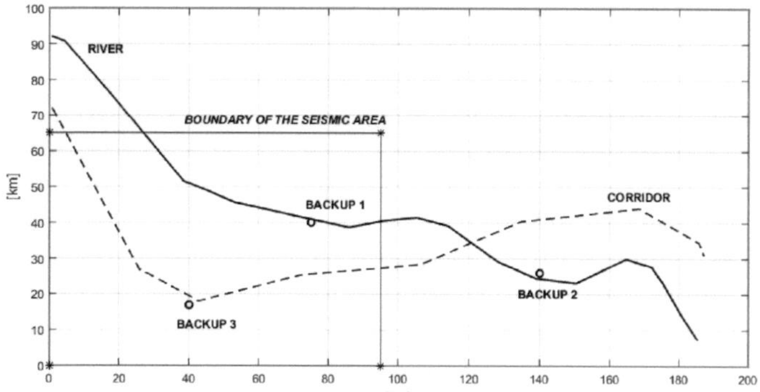

Figure 4.
Spatial layout of the corridor and of the backup sources.

Parameter	Description	Value	Unit
$p_{1,f}$	Probability to involve backup source 1 – flooding	0.5	—
$p_{2,f}$	Probability to involve backup source 2 – flooding	0.5	—
$p_{3,f}$	Probability to involve backup source 3 – flooding	0	—
λ_e	Earthquake damage distance	5	km
λ_f	Flooding damage distance	5	km
s	Seasonal factor (influence of the season on the event)	0	—
$c_{p,c}$	Peak capacity of the corridor	100	J/h
RT	Repair time	1	h
$c_{m,b1}$	Minimum operative margin in capacity – backup source 1	50	J/h
$c_{m,b2}$	Minimum operative margin in capacity – backup source 2	35	J/h
$c_{m,b3}$	Minimum operative margin in capacity – backup source 3	45	J/h
$\alpha_{t,b1}$	Technical availability of the backup source 1	0.95	—
$\alpha_{t,b2}$	Technical availability of the backup source 2	0.95	—
$\alpha_{t,b3}$	Technical availability of the backup source 3	0.95	—
i	Interruptible capacity	0	J/h
e	Energy intensity for the considered commodity	1	€/J
DBE	Magnitude of the design base earthquake	4.8	
DBF	Maximum discharge of the design base flood	2000	m^3/s
R_a	Current risk value	1	€/y
R'_a	Reassessed limit for risk acceptability	0.1	€/y

Table 3.
Values of the main considered parameters.

independent. Furthermore, the values have been set in order to describe a realistic configuration from a physical point of view, while from the economic perspective a unitary value for current risk limit (1 €/y) has been selected mainly due to the unavailability of specific public data on the total economic losses and expenditures. In the reassessment of the limit for risk acceptability, the hypothesis of reducing it by an order of magnitude has been made. In general, if the proposed procedure is applied to a real system, the evaluation of the parameters should be performed according to the considerations expressed in Section 2.1.

The obtained CI (z_c) is shown in **Figure 5** for both earthquake (E) and flooding (F) events. In particular, it can be observed that the corridor sections characterised by the highest CI values are those close to the backup sources in the seismic area (in the case of earthquake event) and to the river (in the case of flooding event). The sections where $CI < 1$ are those corresponding to a damage $D < 0$, i.e. the capacity of the backup sources is more than the one requested to ensure the coverage of the load in the case of unavailability of the corridor.

However, it has to be remarked that all the sections characterised by CI value slightly lower than 1 have to be considered as they are close to a critical condition.

Referring to the evolution of the availability parameter α_b (s,p) for the three backup sources, it can be noticed (**Figure 6**) that the lower the distance between the corridor and the source, the lower the availability: this is because if the natural event involves an area in which the corridor and the backup are close to each other, the probability for the backup source to be damaged is higher, and so its availability is lower.

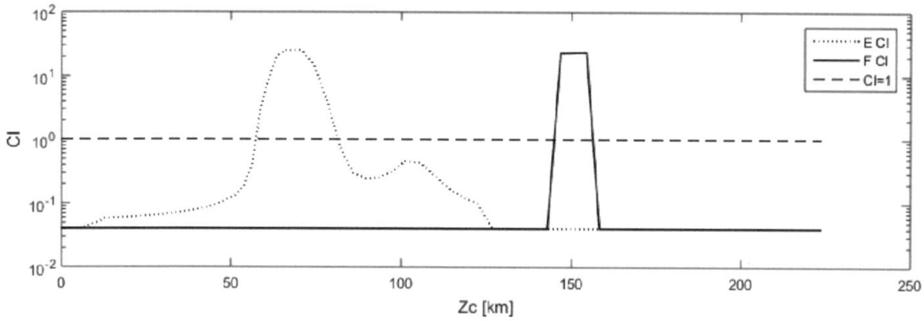

Figure 5.
CI evolution with respect to the position along the corridor z_c*;* CI < 1 *corresponds to* D < 0*.*

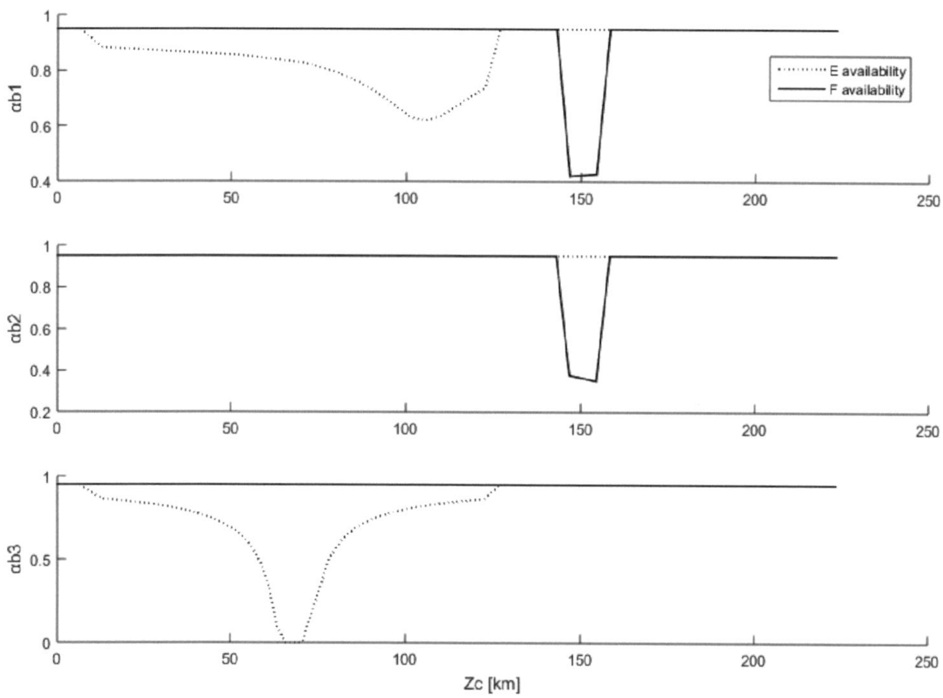

Figure 6.
Evolution of the availability of the backup sources with respect to the position along the corridor z_c*.*

Figure 7(a) shows the frequency-*CI* curves corresponding to the original limit for risk acceptability and to the reassessed one. **Figures 7(b)** and **(c)** represent the frequency-magnitude curves, which have been built by using two different approaches for the two considered classes of natural events:

- the Gutenberg-Richter law [55] in the case of earthquakes;
- a logarithmic relationship based on the one proposed by Wald et al. [56] in the case of flooding.

The vertical lines correspond to the design base earthquake magnitude (DBE) and flood (DBF) for the corridor.

Starting from these curves and from the previously defined *CI* evolution, the maximum acceptable frequencies and the related intensities for both earthquake

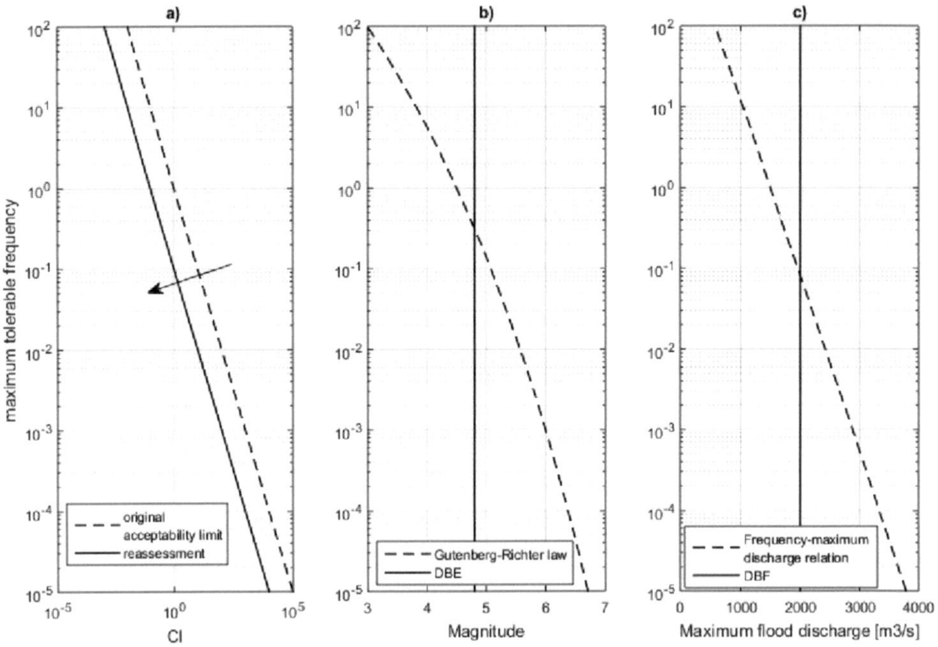

Figure 7.
*Frequency-*CI *(a) and frequency-magnitude curves (b, c) for the analysed case study.*

and flood events and for both the original (E/F old) and reassessed (E/F new) limit for risk acceptability have been estimated, as reported in **Figure 8**.

As it can be observed in **Figure 8a**, the maximum acceptable frequency for earthquakes reaches its minimum value (corresponding to the maximum intensity, visible in **Figure 8b**) in the section where the corridor and the backup source 3 are closest each other and are both affected by the natural event (p = 1 in Eq. (4)). Furthermore, it can be observed that in the case of reassessed risk limit the intensity is beyond the design condition (DBE, **Figure 8b**), thus leading to the need for performing tests in order to assess the robustness of the involved corridor section and to define suitable mitigation actions. The same considerations are valid for the flood (**Figure 8c** and **d**): the main difference is that – in this case – in the most critical corridor section the intensity overcomes the design value also for the original risk limit (DBF, **Figure 8d**), requiring further resilience tests also without hypothesising a reassessment of the limit for risk acceptability.

As mentioned before, the values of the considered parameters have been assumed without a specific reference to a real case, as the goal of the analysed case study is to show the functioning and the applicability of the proposed methodology through a theoretical example. For this reason, an analysis of the uncertainties has not been performed. Future works aiming at deeply exploring the criticality of existing infrastructures will include this aspect, especially regarding the event related parameters, with a particular attention devoted to the probability that different facilities are involved. As previously discussed, in fact, this probability needs detailed and complex considerations to be properly quantified with respect to the specific natural hazard and site studied.

This simplified case study, however, shows the potentiality of this approach in evaluating the possible critical sections of the infrastructures, prioritising the investments and the interventions in reinforcing them and in making them resilient to adverse extreme natural events.

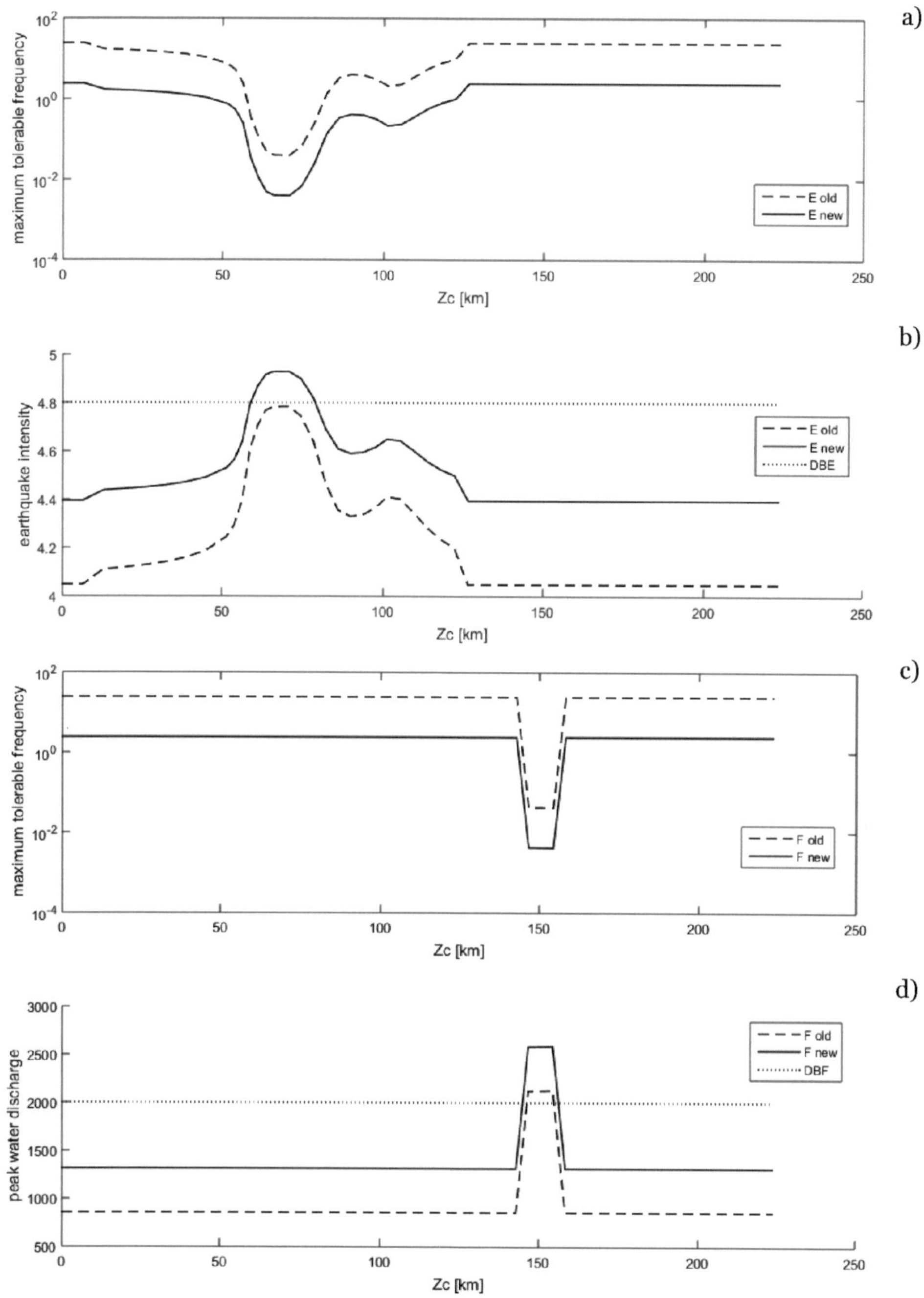

Figure 8.
Maximum tolerable frequencies and intensities of earthquakes (a-b) and floods (c-d) for the analysed cased study.

On the other hand, it also allows to identify some aspects that could be more deeply investigated in future studies in order to enhance the applicability to real cases and the effectiveness of the obtained results. In particular, among them, the unambiguous definition of the system boundaries can be mentioned. In fact, the identification of boundaries can be not easy in the case of meshed networks like natural gas distribution systems or power lines, for which it is difficult to define a single entry point and a single end point. Another relevant aspect is represented by the availability of complete and uniform databases for both the technical characteristics of the analysed infrastructures/backup sources and the classes of natural events affecting the environment surrounding the infrastructure.

4. Conclusions

The protection of Critical Infrastructures against extreme natural hazards by evaluating and improving their resilience is one of the main goals for many countries or groups of countries (like the EU). For this reason, methodologies able to quantify the possible criticalities of these infrastructures are needed to better plan and implement actions, countermeasures and investments allowing to limit or avoid the negative energy, social and economic consequences deriving from natural hazards impacts.

With respect to other studies available in the scientific literature, the approach proposed in this paper focuses on energy corridors and aims at defining a criticality index, which is a function of the spatial position along the analysed corridor, and so it is useful to quantify the criticality level for each section of the considered infrastructure. This index is able to take into account a large variety of parameters (related to the natural event, to the corridor, to the availability of alternative sources and to the involved users) and their interdependencies. The developed methodology can be an effective supporting tool for decision makers and public administrations, for companies that have to manage crucial infrastructures for energy commodities transport and for the civil protection, as it allows – through a simple mathematical formulation – to identify the sections of an energy corridor that are critical with respect to a specific natural hazard or that are close to a criticality status, thus defining priority areas of intervention, preventive investments, mitigation actions and *ad hoc* countermeasures.

The introduced criticality index assesses in a numerical way the socio-economic damage (measured in monetary units) due to the effects of an extreme natural event on the selected infrastructure and can be used to evaluate the maximum acceptable frequency and the corresponding intensity of the event itself, allowing a comparison with the design condition of the corridor.

Furthermore, the possibility to evaluate the criticality index also for negative damage values (i.e. for not critical configurations) permits to measure the distance from the criticality, allowing to pay preventive attention to those sections that are closer to critical situations.

In general, the described approach gives the opportunity of ranking the single branches of a corridor according to their criticality and for all the different natural hazards, and, as a consequence, it gives the authorities in charge of protecting critical infrastructures the opportunity of prioritising the interventions.

The implementation of this methodology on real cases requires specialists from different fields and complex information. This can be deduced also from the application to a simplified case study (considering one corridor and two extreme events). However, the case study has underlined the advantages of the procedure, especially if a reassessment of risk acceptability limit is introduced, because it puts into evidence the safety margin with respect to the design conditions or the need for performing structural tests, quantifying the infrastructure resilience.

Additional aspects should be deeply analysed in the case of an extensive application of the proposed methodology, including – in particular – the availability of complete and homogenous technological and environmental databases and the proper definition of the system boundaries that could be not trivial in the case of meshed networks like the natural gas distribution ones.

Further studies could also be devoted to the analysis of multi-risk scenarios, i.e. to the concurrent occurrence of two or more extreme natural events, defining suitable strategies to allocate the acceptable risk (for instance by taking into account the safety margins of the infrastructure, if they are present), in order to test the infrastructure resilience in the worst (and low-frequency) conceivable conditions.

Author details

Andrea Carpignano[1], Daniele Grosso[2], Raffaella Gerboni[1*] and Andrea Bologna[1]

1 Politecnico di Torino, Torino, Italy

2 LINKS Foundation, EST@Energy Center – Politecnico di Torino, Torino, Italy

*Address all correspondence to: raffaella.gerboni@polito.it

IntechOpen

© 2020 The Author(s). Licensee IntechOpen. This chapter is distributed under the terms of the Creative Commons Attribution License (http://creativecommons.org/licenses/by/3.0), which permits unrestricted use, distribution, and reproduction in any medium, provided the original work is properly cited.

References

[1] European Council. The Stockholm Programme – an open and secure Europe serving and protecting citizens, Official Journal of the European Union, 2010/C 115/01

[2] European Commission. The EU Internal Security Strategy in Action: Five steps towards a more secure Europe, Communication from the Commission to the European Parliament and the Council, COM(2010) 673 final

[3] UNISDR. Terminology on Disaster Risk Reduction [Internet]. 2009. Available from: https://www.undrr.org/publication/2009-unisdr-terminology-disaster-risk-reduction [Accessed: 2020-09-23]

[4] European Commission. Critical Infrastructure Protection in the fight against terrorism, Communication from the Commission to the Council and the European Parliament, COM(2004) 702 final

[5] Commission of the European Communities. Green paper on a European Programme for Critical Infrastructure Protection, COM(2005) 576 final

[6] Commission of the European Communities. Communication from the Commission on a European Programme for Critical Infrastructure Protection, COM(2006) 786 final

[7] The Council of the European Union. Council Directive on the identification and designation of European critical infrastructures and the assessment of the need to improve their protection, Official Journal of the European Union, 2008/114/EC

[8] European Commission. Commission Staff Working Document on a new approach to the European Programme for Critical Infrastructure Protection. Making European Critical Infrastructures more secure, SWD (2013) 318 final

[9] European Commission. Evaluation study of Council Directive 2008/114 on the identification and designation of European critical infrastructures and the assessment of the need to improve their protection, Final Report

[10] European Commission. Commission staff working document. Executive summary of the evaluation of council directive 2008/114 on the identification and designation of European critical infrastructures and the assessment of the need to improve their protection, SWD(2019) 308 final

[11] Ouyang M. Review on modeling and simulation of interdependent critical infrastructure systems. Reliability Engineering & System Safety, 2016;121: 43–60. DOI:10.1016/j.ress.2013.06.040

[12] Griot C. Modelling and simulation for critical infrastructure interdependency assessment: A meta-review for model characterization. International Journal of Critical Infrastructures, 2010;6:363–379. DOI: 10.1016/j.ress.2013.06.040

[13] Wang S, Hong L, Chen X, Zhang J, Yan Y. Review of interdependent infrastructure systems vulnerability analysis. In: Proceedings of the 2nd International Conference on Intelligent Control and Information Processes; 25–28 July 2011; Harbin, China: IEEE; 2011. p. 446–451

[14] Liu W, Song Z. Review of studies on the resilience of urban critical infrastructure networks, Reliability Engineering & System Safety. 2020;193: 1–16. DOI:10.1016/j.ress.2019.106617

[15] Galbusera L, Giannopoulos G, Ward D. Developing stress tests to

improve the resilience of critical infrastructures: a feasibility analysis, Luxembourg: Publications Office of the European Union; 2014. DOI:10.2788/954065

[16] Giannopoulos G, Filippini R, Schimmer M. Risk assessment methodologies for Critical Infrastructure Protection. Part I: A state of the art, Luxembourg: Publications Office of the European Union; 2012. DOI: 10.2788/22260

[17] Haimes YY. Models for risk management of systems of systems. International Journal of System of Systems Engineering. 2008;1:222–236. DOI:10.1504/ijsse.2008.018138

[18] Ariel Pinto C, McShane MK, Bozkurt I. System of systems perspective on risk: towards a unified concept. International Journal of System of Systems Engineering. 2012;3:33.46. DOI: 10.1504/ijsse.2012.046558

[19] Eusgeld I, Nan C, Dietz S. System of systems approach for interdependent critical infrastructures. Reliability Engineering & System Safety. 2011;96:6: 679–686. DOI: 10.1016/j.ress.2010.12.010

[20] Labaka L, Hernantes J, Sarriegi JM. Resilience framework for critical infrastructures: An empirical study in a nuclear plant. Reliability Engineering & System Safety. 2015;141:92–105. DOI: 10.1016/j.ress.2015.03.009

[21] Labaka L, Hernantes J, Sarriegi JM. A holistic framework for building critical infrastructure resilience. Technological Forecasting and Social Change. 2016;103:21–33. DOI: 10.1016/j.techfore.2015.11.005

[22] Mao Q, Li N, Peña-Mora F. Quality function deployment-based framework for improving the resilience of critical infrastructure systems. International Journal of Critical Infrastructure Protection. 2019;26:100304. DOI: 10.1016/j.ijcip.2019.100304

[23] Nan C, Sansavini G. A quantitative method for assessing resilience of interdependent infrastructures. Reliability Engineering & System Safety. 2017;157:35–53. DOI: 10.1016/j.ress.2016.08.013

[24] Ouyang M, Liu C, Xu M. Value of resilience-based solutions on critical infrastructure protection: Comparing with robustness-based solutions. Reliability Engineering and System Safety. 2019;190:106506. DOI: 10.1016/j.ress.2019.106506

[25] Theocharidou M, Giannopoulos G. Risk assessment methodologies for critical infrastructure protection. Part II: A new approach, Luxembourg: Publications Office of the European Union; 2015. DOI:10.2788/621843

[26] DeLaurentis D. Role of humans in complexity of a system-of-systems. In: Duffy VG, editor. Digital Human Modeling. Berlin-Heidelberg: Springer; 2007. p. 363–371. DOI: 10.1007/978-3-540-73321-8

[27] Ferrario E, Pedroni N, Zio E. Evaluation of the robustness of critical infrastructures by Hierarchical Graph representation, clustering and Monte Carlo simulation. Reliability Engineering & System Safety. 2016;155: 78–96. DOI: 10.1016/j.ress.2016.06.007

[28] Kröger W, Zio E. Vulnerable Systems. London: Springer; 2011. DOI: 10.1007/978-0-85729-655-9

[29] Zio E. Critical Infrastructures Vulnerability and Risk Analysis. European Journal for Security Research. 2016;1:97–114. DOI: 10.1007/s41125-016-0004-2

[30] Zio E. Challenges in the vulnerability and risk analysis of critical infrastructures. Reliability Engineering

& System Safety. 2016;152:137–150. DOI: 10.1016/j.ress.2016.02.009

[31] Johansson J, Hassel H, Zio E. Reliability and vulnerability analyses of critical infrastructures: Comparing two approaches in the context of power systems. Reliability Engineering & System Safety. 2013;120, pp. 27–38. DOI: 10.1016/j.ress.2013.02.027

[32] Johansson J, Hassel H. An approach for modelling interdependent infrastructures in the context of vulnerability analysis. Reliability Engineering & System Safety. 2010;95: 1335–1344. DOI: 10.1016/j.ress.2010.06.010

[33] Stergiopoulos G, Kotzanikolaou P, Theocharidou M, Gritzalis D. Risk mitigation strategies for critical infrastructures based on graph centrality analysis. International Journal of Critical Infrastructure Protection. 2015;10:34–44. DOI: 10.1016/j.ijcip.2015.05.003

[34] Kotzanikolaou P, Theoharidou M, Gritzalis D. Assessing n^{th}-order dependencies between critical infrastructures. International Journal of Critical Infrastructures. 2013;9:93–110. DOI: 10.1504/ijcis.2013.051606

[35] Kotzanikolaou P, Theoharidou M, Gritzalis D. Cascading effects of common-cause failures in critical infrastructures. In: Butts J, Shenoi S editors. Critical Infrastructure Protection VII. Heidelberg: Springer; 2013. 171–182. DOI: 10.1007/978-3-642-45330-4

[36] Stergiopoulos G, Kotzanikolaou P, Theocharidou M, Lykou G, Gritzalis D. Time-based critical infrastructure dependency analysis for large-scale and cross-sectoral failures. International Journal of Critical Infrastructure Protection. 2016;12:46–60. DOI: 10.1016/j.ijcip.2015.12.002

[37] Kotzanikolaou P, Theoharidou M, Gritzalis D. Interdependencies between critical infrastructures: Analyzing the risk of cascading effects. In: Bologna S, Hammerli B, Gritzalis D, Wolthusen S. editors. Critical Information Infrastructure Security. Berlin-Heidelberg: Springer-Verlag; 2013. 104–115. DOI: 10.1007/978-3-642-41476-3

[38] Fu G, Khoury M, Dawson R, Bullock S. Vulnerability Analysis of Interdependent Infrastructure Systems. In: Proceedings of the European Conference on Complex Systems (ECCS'12); September 2012; Brussels. Springer; 2012. p. 317–323

[39] Utne IB, Hokstad P, Vatn J. A method for risk modeling of interdependencies in critical infrastructures. Reliability Engineering & System Safety. 2011;96:671–678. DOI: 10.1016/j.ress.2010.12.006

[40] Azzini I, Dido M. GRRASP version 3.1 User Manual. In: Giannopoulos G, Galbusera L, editors. Luxembourg: Publications Office of the European Union; 2016. DOI: 10.2760/999066

[41] Khalil YF. A novel probabilistic timed dynamic model for physical security attack scenarios on critical infrastructures. Process Safety and Environmental Protection. 2016;102: 473–484. DOI: 10.1016/j.psep.2016.05.001

[42] Urlainis A, Shohet IM, Levy R. Probabilistic Risk assessment of Oil and Gas infrastructures for Seismic Extreme Events. Procedia Engineering. 2015;123: 590–598. DOI: 10.1016/j.proeng.2015.10.112

[43] Shakou LM, Wybo J, Reniers G, Boustras G. Developing an innovative framework for enhancing the resilience of critical infrastructure to climate change. Safety Science. 2019;118, 364–378. DOI: 10.1016/j.ssci.2019.05.019

[44] Munich Re. Loss events worldwide 1980–2015 [Internet]. 2016. Available at: https://reliefweb.int/sites/reliefweb.int/files/resources/Loss_events_worldwide_1980-2015.pdf [Accessed: 2020-09-24].

[45] World Bank. World Bank statistical database [Internet]. 2020. Available at: http://data.worldbank.org/ [Accessed 2020-09-24]

[46] Agarwala SK. Principles of Economics. New Delhi; Excel Books India; 2009. p. 324.

[47] Bhattacharyya SC. Energy Economics. Concepts, Issues, Markets and Governance. London: Springer-Verlag; 2011. p. 721. DOI: 10.1007/978-0-85729-268-1

[48] Hungr O, McDougall S, Wise M, Cullen M. Magnitude–frequency relationships of debris flows and debris avalanches in relation to slope relief. Geomorphology. 2008;96:355–365. DOI: 10.1016/j.geomorph.2007.03.020

[49] Jakob M, Friele P. Frequency and magnitude of debris flows on Cheekye River, British Columbia. Geomorphology. 2010;114:382–395. DOI: 10.1016/j.geomorph.2009.08.013

[50] Jakob M, Holm K, McDougall S. Debris-Flow Risk Assessment, Oxford Research Encyclopedia of Natural Hazard Science. Oxford University Press; 2016. DOI: 10.1093/acrefore/9780199389407.013.37

[51] Riley KL, Bendick R, Hyde KD, Gabet EJ. Frequency–magnitude distribution of debris flows compiled from global data, and comparison with post-fire debris flows in the western U.S. Geomorphology. 2013:191;118–128. DOI: 10.1016/j.geomorph.2013.03.008

[52] Hooke JM. Variations in flood magnitude–effect relations and the implications for flood risk assessment and river management. Geomorphology. 2015;251:91–107. DOI: 10.1016/j.geomorph.2015.05.014

[53] Zhang Q, Gu X, Singh VP, Sun P, Chen X, Kong D. Magnitude, frequency and timing of floods in the Tarim River basin, China: Changes, causes and implications. Global and Planetary Change. 2016;139:44–55. DOI: 10.1016/j.gloplacha.2015.10.005

[54] Papadakis G, Vallianatos F, Sammonds P. Non-extensive statistical physics applied to heat flow and the earthquake frequency–magnitude distribution in Greece. Physica A. 2016; 456:135–144. DOI: 10.1016/j.physa.2016.03.022

[55] Gutenberg B, Richter CF. Magnitude and Energy of Earthquakes, Annali di Geofisica, 1956;9:pp. 1–15. DOI: 10.4401/ag-4588

[56] Wald DJ, Jaiswal KS, Marano KD, Bausch D. Earthquake Impact Scale. Natural Hazards Review. 2011;125–139. DOI: 10.1061/(ASCE)NH.1527-6996.0000040

Chapter 4

Resilience and Situational Awareness in Critical Infrastructure Protection: An Indicator-Based Approach

Aleksandar S. Jovanovic, Somik Chakravarty and Marjan Jelic

Abstract

The paper proposes a concept enabling quantitative assessment of resilience in critical entities developed in the European projects SmartResilience and InfraStress. The concept aims at combining simple communication-related advantages of simplified assessments results (such as "resilience very high" or "resilience very low") with the advantages of the in-depth assessments (e.g. analysis of multiple sensor data). The paper describes the main elements of the innovative, indicator-based concept, starting with the "resilience cube" at the top, and continuing with the multi-level, hierarchical, indicator-based assessment methodology. The concept allows analyzing and assessing different aspects of practical resilience management. One can assess the resilience level of an entity at a given point in time, monitor their resilience level over time and benchmark it. One can also model and analyze the functionality of a system during a particular (threat) scenario, as well as stress-test it. The same methodology allows to optimize investment in improving resilience (e.g. in further training, in equipment, etc.), in a transparent and intuitive way. A resilience indicator database (over 4,000 indicators available) and a suite of tools (primarily developed within SmartResilience and InfraStress projects) and a repository of over 20 application cases and 300 scenarios, support application of the methodology. The concept has been discussed and agreed with over 50 different organizational stakeholders and is being embedded into the new ISO 31050 standard currently under development. Its "life-after-the-project" will be ensured by the dedicated "resilience rating initiative (ERRA)". Although the concept and the tool in the form of the "ResilienceTool" were developed primarily for the resilience assessment of critical infrastructure (the "smart" ones in particular), they can be used for resilience assessment of other systems and through the extension of the, already initiated, implementation of AI techniques (machine learning) to make the ResilienceTool even more versatile and easier to use in the future.

Keywords: resilience, risk assessment, critical infrastructure, resilience indicators, risk and resilience

1. Introduction: using indicators to assess and manage resilience of critical infrastructures in SmartResilience and InfraStress projects

Modern critical infrastructures are becoming increasingly smarter (e.g. the smart cities). Making the infrastructures smarter usually means making them smarter in the normal operation and use: more adaptive, more intelligent etc. But will these smart critical infrastructures behave smartly and be smartly resilient also when exposed to extreme threats, such as extreme weather disasters or terrorist attacks? If making an existing infrastructure smarter is achieved by making it more complex, would it also make it more vulnerable? Which aspect of resilience of a critical infrastructure will be affected the most? Its ability to anticipate, to prepare for, to adapt and withstand, respond to, or to recover? What are the resilience indicators (RIs) which one has to look at? These are the main questions tackled by the SmartResilience project [1] to which a methodology based on resilience indicators was developed, complete with the supporting "ResilienceTool" to handle both existing ("conventional") indicators suitable for assessing the resilience of critical infrastructure as well as new "smart" resilience indicators, e.g. those from Big Data (over 5,000 available in mid-2020). In the InfraStress project [2], the concept and the tools are developed further and integrated with the concept of situational awareness system (focus of the InfraStress project).

2. Resilience as "one number", ResilienceCube and the main concept

2.1 Resilience and resilience matrix

The definition of resilience, standing in the background of the concept presented in this paper, has evolved along with the work on the development of the concept. It started with the definition of the resilience as *"The ability to anticipate, prepare for, and adapt to changing conditions and withstand, respond to, and recover rapidly from disruption".* The main amendment proposed afterward was the inclusion of the ability to understand risks (current and emerging), leading to the definition of *"Resilience as the ability to understand risks, anticipate, prepare for, and adapt to changing conditions and withstand, respond to, and recover rapidly from disruption"* [3]. In the final stage, the project adopted the elaborated definition of the resilience of an infrastructure is given below [4].

"Resilience of an infrastructure is the ability to understand and anticipate the risks – including new/emerging risks – threatening the critical functionality of the infrastructure, prepare for anticipated or unexpected disruptive events, optimally absorb/withstand their impacts, respond and recover from them, and adapt/transform the infrastructure or its operation based on lessons learned, thus improving the infrastructure anti-fragility."

This definition enabled the following main advantages:

- Including emerging risks and a natural link to risk assessment
- Including the goals of optimization, adaptation and transformation and
- Including the improvement of anti-fragility, the concept of increased importance for all smart systems, including smart infrastructures, and
- Enabling inclusion of the 5 phases of the resilience cycle and the indicator-based approach within the resilience matrix.

The definition allows analyzing the behavior of an infrastructure exposed to an adverse event over a "scenario timeline" and simultaneously assessing the functionality of an infrastructure over the "resilience cycle" as shown in **Figure 1**. While the decomposition over the time-axis, i.e., defining the "phases" of the resilience cycle, may be trivial, decomposition over the functionality axis is non-trivial as functionality might have different "dimensions" (see chapter 2.3). The SmartResilience concept proposes the decomposition over a 5 × 5 resilience matrix, defining 5 phases and 5 "dimensions".

The approach allows to represent the overall resilience cycle, and focus on single relevant issues. The issues, in turn, can be described by means of indicators and these can have values, thus, providing the possibility to quantitatively describe each "cell" of the resilience matrix (**Figure 2**).

Phase I, understand risks, is applicable prior to an adverse event. It emphasizes emerging risks and includes their early identification and monitoring; e.g. what could the "adverse event" be? This is followed by.

Phase II, anticipate/prepare, also applicable before the occurrence of an adverse event. It includes planning and proactive adaptation strategies, possibly also "smartness in preparation" [5].

Phase III, absorb/withstand, comes into action during the initial phase of the event and shall include the vulnerability analysis and the possible cascading/ripple effects; e.g. "how steep" is the absorption curve, and "how deep" down will it go?

Phase IV, respond/ recover, is related to getting the adverse event under control as soon as possible, influencing the "how long" will it last, question. Further, it

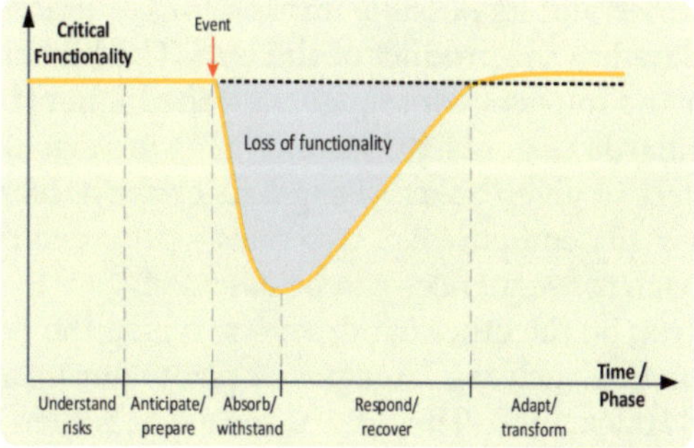

Figure 1.
The 5 × 5 resilience matrix, mapping the critical infrastructure system functionality over 5 phases of the resilience cycle.

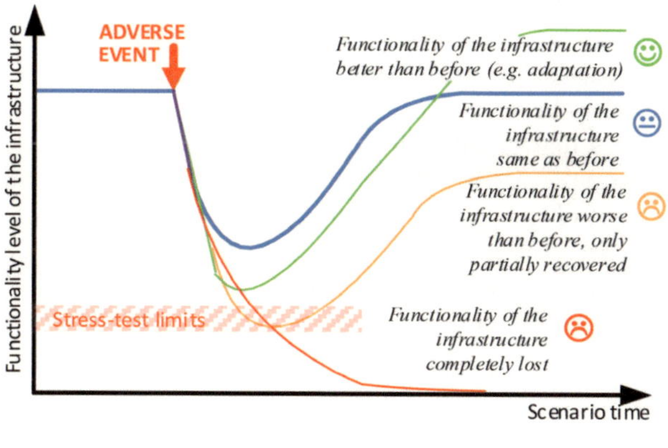

Figure 2.
Possible outcomes of case of an infrastructure exposed to an adverse event: Between improvement and complete failure.

includes the post event recovery; e.g. "how steep up" is the recovery curve for normalization of the functionality? It is followed by.

Phase V, adapt/learn, which encompass all kinds of improvements made on the infrastructure and its environment; e.g. affecting "how well" the infrastructure is adapted after the event, and whether it is more resilient and "sustainable". The activities in this phase also lead to preparation for future events and hence, this resilience curve also exhibits a reoccurring cycle [5].

The dimensions help in categorizing the indicators. The system/physical dimension includes technological aspects, as well as the physical/technical networks being part of a given infrastructure, and the interconnectedness with other infrastructures and systems. The information/data dimension is related to the technical systems. The organizational/business dimension covers business-related aspects, financial and HR aspects as well as different types of respective organizational networks. The societal/political dimension encompasses broader societal and social contexts. Finally, the cognitive/decision-making dimension, accounts for perception aspects (e.g. perceptions of threats and vulnerabilities) [6].

2.2 Difference and relationship between a risk matrix and a resilience matrix

One should distinguish well between the risk matrix and the resilience matrix. Although similar in shape and appearance, their basic purpose and principles are different. The main purpose of a risk matrix is to show the position of a given risk (defined through its scenario) on a 2-dimensional "map", depicting the likelihood/probability of a given risk and its possible impact/consequences. Risk is then, for a given scenario, calculated as the product of the two. The higher the probability/likelihood, the higher the impacts/consequences – the higher the risk.

Risk-oriented standards (e.g. EN 16991:2018[1] [7]) provide detailed examples of how to use a risk matrix in given areas. Using a risk matrix (sometimes referred as "risk map"), one can easily compare e.g. two risks – provided that the likelihood/probabilities and impact/consequences can be assessed.

The resilience matrix, on the other hand, serves to map the resilience of a system (e.g. a critical infrastructure such as a large power plant) during an adverse event (e.g. crisis, accident, cyber-attack, etc.). The time of the event is then usually subdivided into phases (**Figure 3(a)**), usually 4 or 5, of the event, from the time before the very event to the time after the event (the "resilience cycle"). The time of the event/scenario (see also **Figure 4**) is thus, the first and the main dimension of the resilience matrix. As the adverse event, in a general case, will affect different areas of activities, e.g. business, society, information, management, etc. the event is usually looked at for each of them in terms of their own indicators. These areas are often (e.g. in EU projects such as InfraStress [2] or SmartResilience [1, 8–12]), called dimensions, and their number is usually chosen as equal to the number of phases. The result is then a matrix (the "resilience matrix", **Figure 3(a)**), mapping the resilience of the given system – e.g. suggesting the communication "dimension" in the response "phase" of the crisis management of COVID (e.g. in the UK[2]) was "poor" (**Figure 3(b)**).

In the approach presented here, we propose that the qualifier "poor" is linked to the measurable indicators (resilience indicators) such as e.g. reliability of numbers communicated to the public, statistic/sentiment in social media, survey results, etc. In such a case, the label "poor" is supported also by quantitative indicators and can be given an aggregated value (e.g. acc. to the value × weight formula).

---

[1] https://www.cen.eu/news/brief-news/pages/news-2018-011.aspx (Convener A. Jovanovic).

[2] https://reutersinstitute.politics.ox.ac.uk/communications-coronavirus-crisis-lessons-second-wave

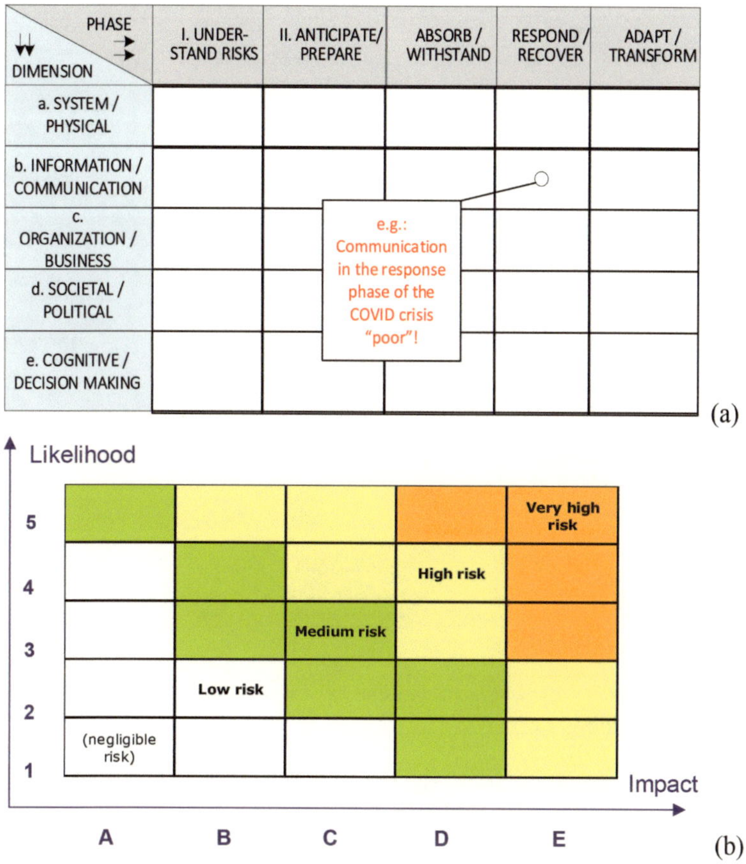

Figure 3.
Example of a 5x5 resilience matrix (a) as compared to a risk matrix (b).

Generally, the aggregation process for indicators in the method and the tool described (see **Figure 5**) here offer the following main aggregation options:

1. The simple aggregation of the indicators put on the common 0–5 scale

2. The weighted aggregation as an extension of the simple method

3. The JRC composite indicators and scoreboard (COIN) methodology[3]

4. The Fuzzy-AHP based weight determination [13]

5. The ranking-based weight determination [11]

2.3 "Measuring" resilience by means of issues and indicators

In the concept, an "issue" is a general term referring to anything important in order to be resilient against severe threats such as terror attacks, cyber threats and extreme weather. It is telling what is important, e.g., it can be "training" performed in the anticipate/prepare phase. Obviously, the more indicators one chooses, the better the "coverage" of an issue is going to be (**Figure 5**), but it is also obvious that

[3] https://ec.europa.eu/jrc/en/publication/coin-tool-user-guide

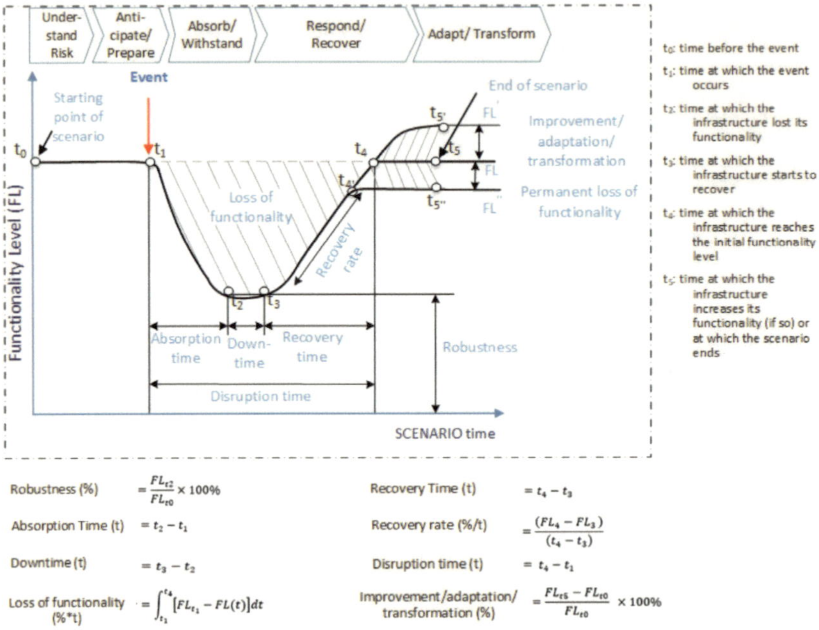

Figure 4.
Functionality level of the smart critical infrastructure over scenario time – The value of the FL at a particular time is calculated by aggregating the relevant indicators scores starting from FL at t_0 = *100%.*

the larger the number of indicators, the more complex their handling is going to be. The "way out" has two components and these would be:

- finding the "right number" of indicators acc. to the resilience problem tackled (in the usual engineering practice, managed by humans, 120–150 indicators are usually a maximum – the more critical the situation, the smaller the number; in absolute emergency situations humans can hardly look at more than 3 indicators), and

- allowing to "drill-down" in cases when one or more indicators need further explanation.

In order to organize the analysis and enable drilling down to the base assessment elements, the selected scenario is segmented into six levels [1]. This practice is based on several previous methods, notably the ANL/Argonne method [14], the Leading Indicators of Organizational Health (LIOH) method [15–17], the US-DHS method [18], and the Resilience-based Early Warning Indicator (REWI) method [19]. The ANL/Argonne method for assessing a resilience index (RI) is structured in 5 levels, providing indicators on the lowest level and a similar hierarchy is used in the SmartResilience and InfraStress projects for assessing resilience levels, entering the indicators on level six.

The "resilience indicators" are mainly taken from current practices (standards, guidelines, reports, etc.) within safety and risk management, emergency preparedness, business continuity, etc. and in most cases, they exist already as safety indicators, risk indicators, or similar (e.g. those proposed by OECD, GRI, API, HSE, IAEA and other organizations). Collecting the indicators and applying the approach, the theoretical framework for variable selection, weighting, and aggregation must be defined [20] and the basis for this is the context of the assessment, or scenario. An example of a "resilience indicator data sheet" is given Appendix 1.

The values of indicators, often for one and the same indicators, can come from experts (e.g. as qualifiers – "high", "very low"", etc.), from measured or monitored

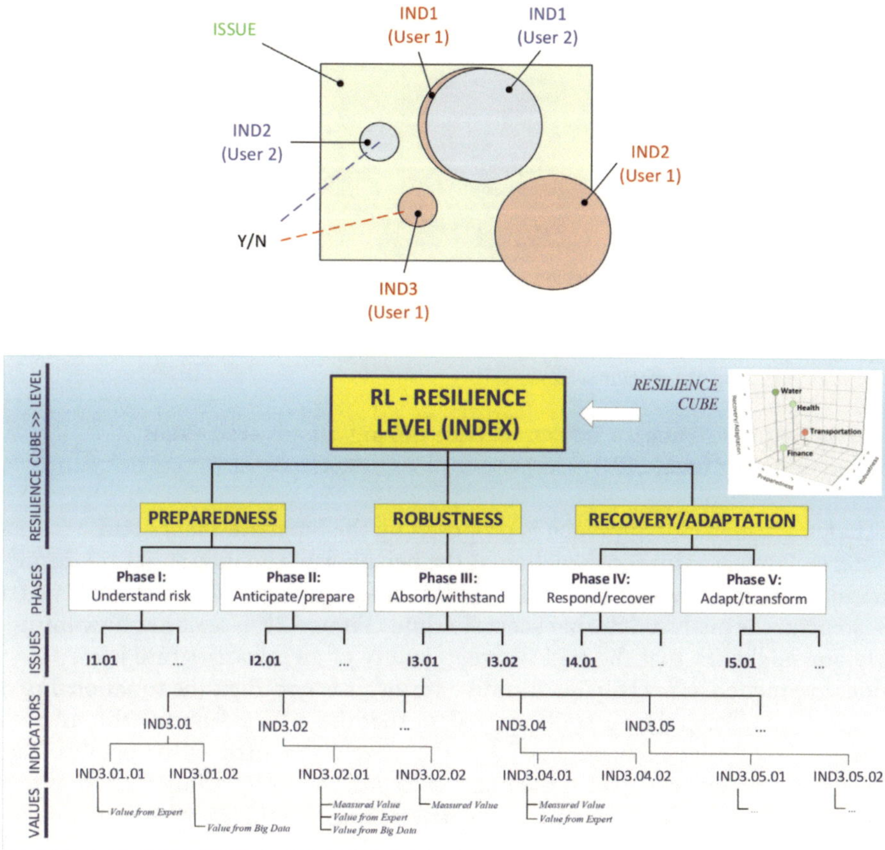

Figure 5.
Issues measured by indicators (above), allow to make the bridge between a given, e.g. measured value of an indicator, and the overall, final resilience index & ResilienceCube (below).

values (e.g. numbers of accidents), or from big data analysis. Single, real values, from any of the above sources, in the methodology, can be yes/no questions, numbers, percentages, fuzzy numbers, or some other type. Once in the model, for the communication with the end-user, they are, in a general case, transferred into the score, on a scale 0 to 5.

2.4 Dynamic checklists of resilience issues and indicators

One of the ways to use resilience issues and indicators practically [21], is to put them into "lists" (checklist) and in the concept it is done in a dynamic way, allowing to dynamically create checklist appropriate for a given case using available indicators or adding new ones to the list. In order to make the creation/drafting of these dynamic checklists (DCLs) easier and allow for comparison and benchmarking of results, the user is encouraged to use the list suggested by the concept, namely (**Figure 6**):

- The CORE DCLs, containing the indicators suggested for virtually all infrastructures,
- The RECOMMENDED DCLs, containing indicators suggested for the particular type of infrastructures and
- The USER's DCL, containing indicators specific for a particular infrastructure.

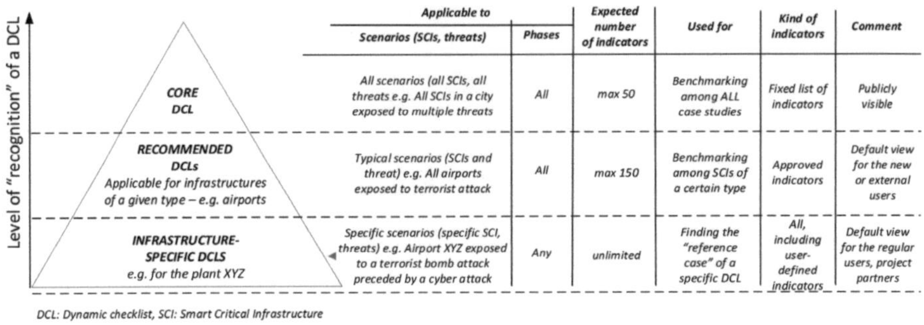

Figure 6.
Hierarchical structure of the checklist in the concept.

2.5 Assessing resilience an infrastructure during an adverse event: Functionality level (FL)

The indicator-based approach is proposed by the SmartResilience and InfraStress projects also for modeling of the behavior of the infrastructure during a particular disruptive event (scenario). In this case, the (critical) functionality of an infrastructure is analyzed during scenario time (**Figure 2**). No matter how intuitively one might say that the critical functionality of an infrastructure is easy to define, in practice, especially quantitative terms, it is not. E.g., the functionality of an airport is to "keep the air traffic going" or that the critical functionality of a refinery is "to produce the gasoline", but these are often difficult to measure. E.g., in the air traffic, one can look at the number of passengers boarding and/or on cargo throughput, but should at the same time look at the compliance with, e.g., safety and environmental norms, because not satisfying the latter could also be a loss of critical functionality. In the concept, these are considered to be

- The ELEMENTS of the functionality (corresponding to the "issues"), and for this one can define
- The (FUNCTIONALITY) INDICATORS, just as in the case of resilience level assessment.

Defining the functionality in the above way enables to precisely and quantitatively define the resilience curve in scenario time, e.g. for the main characteristic points in time [22]:

t_0: time before the event or starting point of the scenario.

t_1: time at which the event occurs.

t_2: time at which the infrastructure reaches the minimum functionality level.

t_3: time at which the infrastructure starts to recover.

t_4: time at which the infrastructure reaches the initial functionality level or starting point of a new steady-state level.

t_5: time at which the infrastructure increases its functionality through learning and adapting or at which the scenario ends.

Based on the resilience curve (or functionality curve), it is then possible to define the resulting macro-indicators, as illustrated in the notional diagram in **Figure 4**, such as:

- Robustness [%]
- Absorption time [h]

- Downtime [h]
- Loss of functionality [% over h]
- Recovery time [h]
- Recovery rate [%]
- Disruption time [h]
- Improvement/adaptation/transformation [%]

It should be noted that these are the RESULTING macro-indicators, and not the INPUT indicators as the resilience indicators and functional indicators mentioned above. These macro-indicators can also be used for "stress-testing", in which case these can be compared with the critical thresholds (e.g. for the maximum loss of functionality, duration or a combination of these, etc.).

Robustness characterizes the absorbing capacity of the smart critical infrastructure [23]. NL uses robustness as defined by the National Infrastructure Advisory Council (NIAC) [24], i.e. "the ability to maintain critical operations and functions in the face of crisis" [25]. It can be seen as the protection and preparation of a system facing a specific danger. The objective of the robustness component is to identify measures that can help the system withstand or adapt to a hazard. It emphasizes the ability of an infrastructure to withstand the incident if the protective measures fail. It also integrates the capacity of the infrastructure to function in a degraded state. The importance of robustness is not necessarily defined by how the infrastructure continues to function in the face of an incident but rather by how it is able to continue to accomplish its mission and to provide its products and services through preventative measures, mitigation, or absorption capabilities [25]. Robustness is defined as the capacity of the smart critical infrastructure to endure the effects of a negative event and thereby absorb its impact. As shown in **Figure 4**, it is measured as the ratio of the percentage of the lowest FL after the disruption, i.e. at time t_2, to the FL during normal operation, i.e. at time t_0.

$$\text{Robustness} = \frac{FL_{t2}}{FL_{t0}} \times 100\% \tag{1}$$

Absorption time is defined as the time during which the smart critical infrastructure absorbs a disruptive event while the smart critical infrastructure undergoes a decrease in its functionality level. As illustrated in **Figure 4**, it is measured as the difference between t_2 and t_1.

$$\text{Absorption time} = t_2 - t_1 \tag{2}$$

Loss of functionality is the functionality of the smart critical infrastructure lost in a given threat situation. It is measured by the area of the curve (an approximation) between the time when the smart critical infrastructure starts to lose its functionality (t_1) to the time when it reaches the initial state (t_4) (see **Figure 4**). The approximation is done for the area above the curve to a well-defined shape, e.g. a triangle. The output would be the percentage loss of functionality in time [26, 27], e.g. losing 10% in 10 hours.

$$\text{Loss of functionality} = \int_{t_1}^{t_4} [FL_{t_1} - FL(t)]dt \tag{3}$$

FL in all the formulae (incl. Eq. (3), is calculated as the aggregated score on indicators, in the particular case of FL, as functionality indicators, such as those presented in the sample list in **Figure** 7).

Next in the scenario is the **recovery** state of the smart critical infrastructure. The concept of recovery explains the passage of an infrastructure's functionality from a degraded state to one of acceptable operation. This concept builds on the concept of robustness in that, if measures of robustness fail to fully prevent, mitigate, or allow the asset to absorb the damage event, recovery constrains the impacts of the event to keep the CI functional. For the purpose of modeling the impact of a disruptive event, **recovery** refers to the ability to not only return to acceptable operating levels but also to recover fully from the effects of an event [25] in the maximum allowable/acceptable recovery time (as described in the stress test methodology [12, 28]).

Downtime is defined as the time duration for which the system is not functional. In Ref. to critical infrastructures, this could apply if the CI stops functioning. In this case, the functionality level of the infrastructure remains below the **threshold level** of functionality [25]. It can be measured as the difference in time between t_3 and t_2 (see **Figure 4**).

$$\text{Downtime} = t_3 - t_2 \tag{4}$$

Note: This calculation is conducted when the threshold level of functionality is defined (Here it is assumed that the threshold level is FL_{t2} ($=FL_{t3}$)).

Recovery time is defined as the time at which the smart critical infrastructure recovers from the disruptive event and gains its initial or desired functionality [23]. It can be measured as the time taken to recover the functionality level, i.e. the time between time t_3 and t_4.

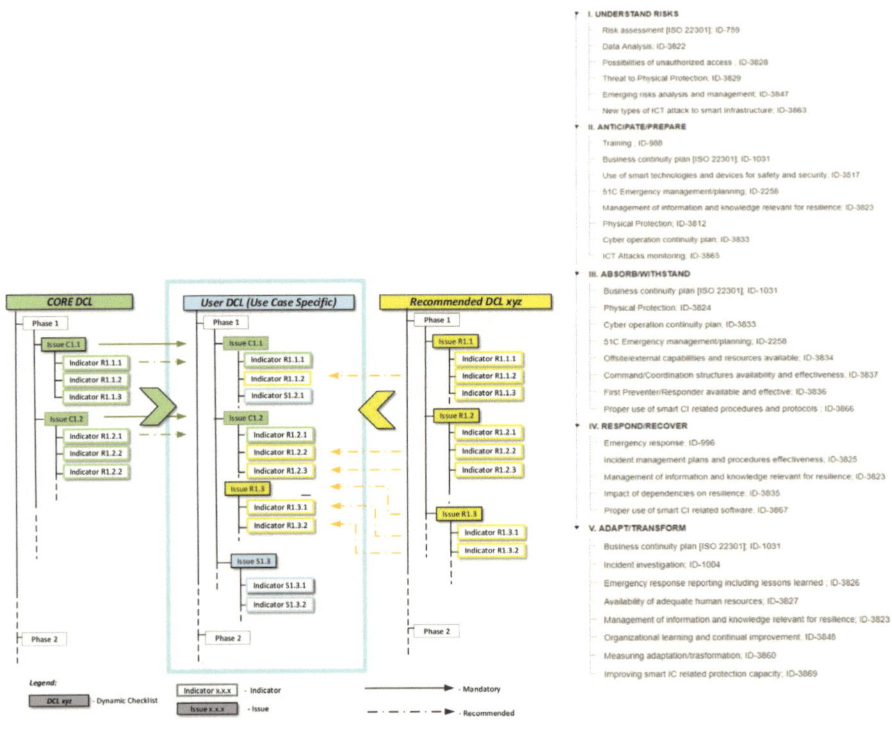

Figure 7.
Example of creating a DCL by combining generic (CORE DCL), typical (RECOMMENDED DCL) and specific issues/indicators into the final DCL.

$$\text{Recovery time} = t_4 - t_3 \tag{5}$$

Note: Since the functionality level at the end of the scenario may be different from at the start of the scenario, the recovery time may have to be measured at a new steady-state level [28].

Recovery rate is defined as the rate at which the smart critical infrastructure recovers from a disruptive event and gets back to its initial functionality level [23]. It characterizes the recovery trajectories of the smart critical infrastructure from the point it starts recovering from the scenario to the final recovery. It is measured as the ratio of change in functionality level between time t_3 and t_4.

$$\text{Recovery rate} = \frac{(FL_4 - FL_3)}{(t_4 - t_3)} \tag{6}$$

Another measure considered for modeling the impact is **disruption time.** The disruption time is defined as the total time taken by the CI to recover. It is also seen as a measure for recover capacity of the smart critical infrastructure to return to the desired functionality level [23]. In the functionality level over time (FL-t) curve, it is the time between when the event occurs, i.e. at time t_1, and time when the smart critical infrastructure has fully recovered, i.e. t_4 (see **Figure 4**).

$$\text{Disruption time} = t_4 - t_1 \tag{7}$$

Improvement/adaptation/transformation: Final recovery of the FL of a smart critical infrastructure could be equal to, better than, or worse than the original FL [29]. Hence, the model allows for the calculation of the "improvement/adaptation/

Number of indicators per cell

Phase Dim.	I.Understand Risk	II.Anticipate/ prepare	III.Absorb/ withstand	IV.Respond/ recover	V.Adapt/ transform
a. System/ physical	1	0	4	4	0
b. Information/ smartness	0	3	1	0	0
c. Organization/ business	3	3	1	1	1

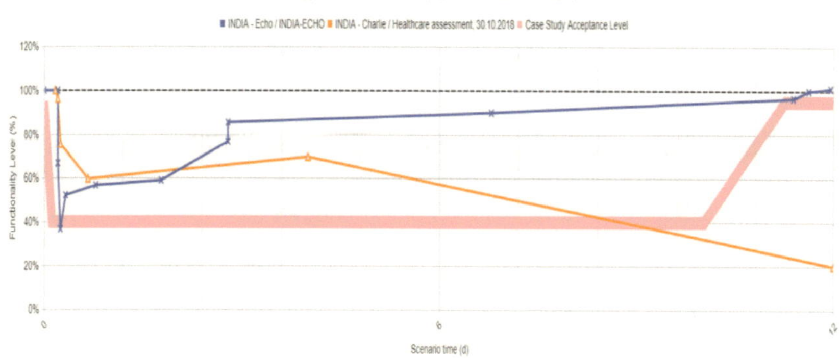

Figure 8.
An example of a report of one of the resilience assessments – FL curve comparing the response of FL with scenario time for case studies ECHO and CHARLIE, including the comparison of the FL curves with the acceptance level (shown in pink, can be used for stress-testing, too).

Macro Indicator	Values for ECHO	Values for CHARLIE
Robustness [%]	42	20
Absorption time [h]	1	284
Downtime [h]	2	−192
Loss of functionality [% over h]	58% in 282 h	80% in 284 h
Recovery time [h]	279	192
Recovery rate [%]	17	−26
Disruption time [h]	0	284

Table 1.
*The macro indicator values for the cases in **Figure 8** - the macro indicators calculated from the FL curve provide a quantitative way of comparing alternatives of system recovery supporting decision making and optimization [1].*

transformation." This is the capacity of the smart critical infrastructure to learn from a disruptive event (e.g. a revision of plans, modification of procedures, introduction of new tools and technologies [10]) (see **Figure 4**). It is measured as the ratio of change in FL during and after the event over the initial FL.

$$\text{Improvement/adaptation/transformation} = \frac{FL_{t5} - FL_{t0}}{FL_{t0}} \times 100\% \tag{8}$$

Such macro indicators are ideal for comparing the FL responses for multiple case studies, infrastructure, entities etc. They allow an objective evaluation of not only how the functionality level of a system might react to an event but also how and when it can recover. Using a theoretical acceptance level, a stress-test can also be performed. An illustrative example comparing the FL response for two SmartResilience case studies (ECHO and CHARLIE) is shown in **Figure 8** and **Table 1**.

3. Practical application of the ResilienceCube and the methodology for resilience assessment

The indicator-based resilience concept described above, enables practical assessment of the following aspects of resilience (**Figure 9**):

1. Resilience Index (Resilience as "one number") and the ResilienceCube (preparedness, robustness, adaptation/transformation)

2. Assessing resilience of an infrastructures over time – the Resilience Level (RL)

3. Assessing resilience of an infrastructure during an adverse event – the Functionality Level (FL)

4. Assessing resilience of "multiple infrastructures": Multi-level resilience assessment

5. Modeling interaction and dependencies, visualizing resilience

6. Comparing resilience of different infrastructures: Benchmarking

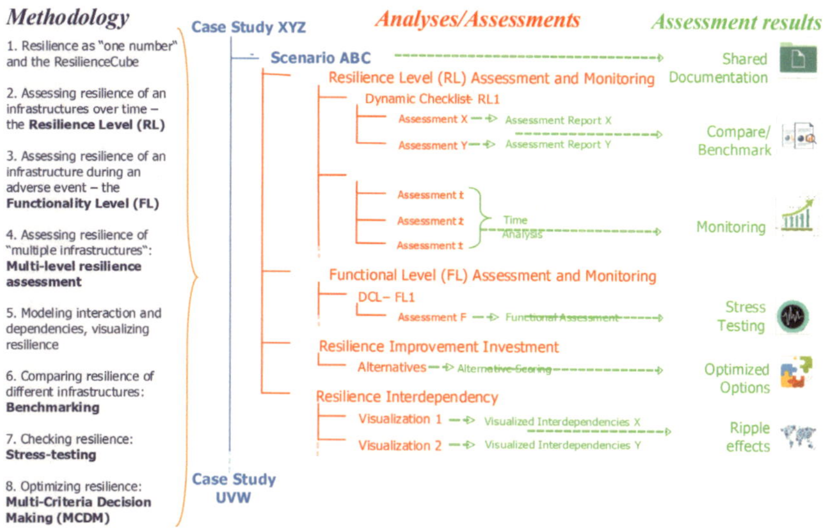

Figure 9.
Applying the methodologies in order to assess resilience and obtain practical (quantitative) results.

7. Checking resilience: Stress-testing

8. Optimizing resilience: Multi-Criteria Decision Making (MCDM)

For its users, the methodologies are embedded into the interactive, web-based and freely available "ResilienceTool". Applied in different case studies, dealing with energy, transportation, health, smart cities, water, sensitive installations, etc., the methodology and tool provide the user with different options when using the approach and the system by showing how benchmarking can be done and the best-practice solutions can be re-used.

When applying the concept and the methodologies practically, it is important to understand that the flexibility of the concept and the methodologies necessarily demand for domain expertise in "configuring" the resilience model for a specific area/city or critical infrastructure. A fixed list of critical infrastructures for cities in Europe does not exist, and it must be up to each user of the concept, methodologies and the software tool, to decide which feature of respective infrastructures should be analyzed and how. Similarly, no fixed list of threats exists, neither on the area level nor for the single critical infrastructures. Thus, it will be up to the users to define which threats (scenarios) they consider relevant. Domain experts are needed in order to define the important issues, and how to measure these issues, i.e. identifying the indicators. They are in a way "configuring" the resilience model, which largely is a one-time effort prior to using the model for calculating the resilience levels, although some adjustments, tuning, and reconsiderations are expected. Thus, in the implementation phase, it is important to have close collaboration between the users, the method developers, and IT developers (of calculation and presentation tools).

3.1 Resilience index/cube, resilience level (RL), functionality level (FL) and multi-level resilience assessment

Per default, assessing resilience in the concept is based on scoring (other ways of upwards aggregation are possible, but used only in "expert mode"), the scores being aggregated upwards – up to the Resilience Index score. At each level, the scores can be assigned weights, as the indicators, too. When performing the resilience assessment, the indicators' real values are entered into the calculation, and the issue scores are obtained as average weighted scores of the indicator scores. It is possible to let a specific indicator overrule the effect of the other indicators, i.e. having "knock out indicators" where, in the case of a low value, the effect is not "averaged away" through an average weighted score of all the indicators. The reasoning behind the selected scales is that a scale from 0 to 5 for indicators (and issues) are sufficiently broad, especially if there are needs to perform expert judgments to provide scores for the indicators (or directly for the issues) in case of lack of data [17]. This has similarities to the use of safety integrity levels (SIL) for safety-instrumented systems [30]. In and for the cases where the issue-indicator approach is not sufficient, the concept and the tool allow using multi-level indicators (de facto composite indicators).

3.2 Modeling interaction and dependencies, visualizing resilience

SmartResilience and InfraStress projects look at interdependencies between infrastructures to understand how, in a case of a problem in one of them, the functionality of others can be impacted. The assessment is based on issues and indicators: these issues and indicators that are shared by different infrastructures indicate "lines of interconnectedness and interdependency". The infrastructures involved and the issues/indicators form thus the logical network that can analyze in order to model the propagation of influences from one infrastructure to another. Thus, the cascading and ripple effects can be modeled and the dynamic behavior of the network ("infrastructure-of-infrastructures") analyzed **Figure 10**).

The network in **Figure 10** is created as the case applied onto the indicators applicable to six types of infrastructures in SmartResilience project (health, ICT, energy, water, transportation, industry) and looking at the core, recommended and specific indicators (**Figure 6**). About 2,000 indicators were considered. The

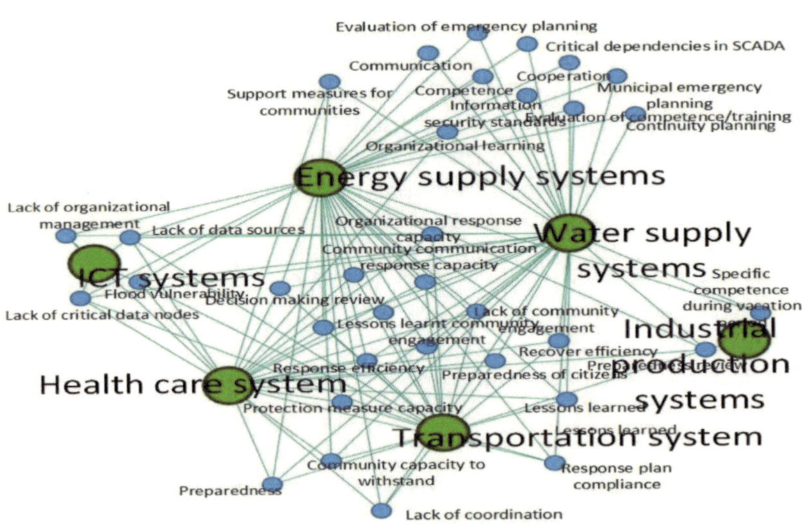

Figure 10.
Interdependencies among multiple infrastructures as a network: Common indicators define the interdependencies.

analysis has included the web-semantics-based analysis of the descriptions of indicators and the statistical analysis of the values of these indicators in the case studies performed in SmartResilience project. The analysis has also served as the basis for the,more user-oriented visualization of interdependencies in a critical infrastructure.

3.3 Comparing resilience of different infrastructures: benchmarking

Using issues and indicators from pre-approved and standardized sources such as the CORE and Recommended DCLs allows for the additional benefit of benchmarking certain aspects of resilience management across different organizations. As the CORE issues are expected to be present in every Complete DCL, organizations can at the very least be compared based on managerial, resilience-oriented activities and processes, regardless of industry or threat. WITHIN a particular scenario (industry and threat), Complete DCLs can be benchmarked when using the Recommended issues proposed by the industry's experts.

Once the CORE DCL issues are selected, the user can make an actual resilience assessment adding the indicators under the CORE issues. Since for all of the case studies, the Recommended DCLs have been developed, one can take a look at those lists and choose which indicators from there fit into the CORE DCL. It may happen that the names of the issues from Recommended DCL are slightly different from the CORE ones. Hence, it is possible that not all the previously used indicators will fit. In this case, the user should use only the ones which match with the CORE issue. Furthermore, it may be needed that new indicators (not used in the Recommended DCL) are added in order to ensure sufficient coverage of the CORE issue.

3.4 Checking resilience: stress testing

The stress test framework is used to test whether, in a given threat situation, the smart critical infrastructure is/will be resilient enough to be able to continue functioning within the prescribed limits. The FL curve(s) obtained in the analysis is compared with the stress test criteria and limits in order to evaluate whether the smart critical infrastructure has passed or failed the stress test. In order to do the stress test, the user needs to decide on the thresholds/limits representing acceptable/non-acceptable values for each criterion. The stress test criteria can be related to (e.g.):

- Functionality Level
- Time (to absorb, to recover)
- Cumulative loss of functionality (area)

Functionality Level ("vertical loss"): the stress test limits can be set based on the overall functionality level, at single functionality element(s), and/or at single functionality indicator(s). The limit could be a certain minimum level of functionality (i.e. the lowest point of the resilience curve should be above this FL_{min}). The functionality level at the lowest point below the curve is sometimes referred to as "robustness," which can be set as a stress test limit.

Time ("horizontal loss"): when subjected to a threat/event, a smart critical infrastructure may set the limits on time (e.g. maximum time to absorb the event, maximum time to partially recover after the event, or maximum time to fully recover after the event). The last time interval, i.e. time between when the event occurs and the smart critical infrastructure is fully recovered, is referred to as disruption time when modeling the impact of a disruptive event. This is sometimes also referred to as "rapidity" and can typically be used as a stress test limit. For example, the stress test limit could be the time from when the event occurs until 90% of the functionality is restored, or some combination of various criteria.

3.5 Optimizing resilience: Multi-criteria decision making (MCDM)

Given that the purpose of the resilience and functionality level assessments is to reveal weaknesses, either isolated or in comparison with others (benchmarking), implementation of improvement measures is expected to be required. Which improvement measure(s) will be optimal to choose? Given a set of alternatives/options various criteria need to be weighed against each other. This could typically include the effect on resilience (e.g. higher RL), costs and time to implement the measure(s), but also other criteria may be relevant. The method used to decide on optimal improvement measures is a Multi-Criteria Decision Making (MCDM) method and given that the nature of smart critical infrastructures and the resilience issues that they evoke tend to mix both quantitative (budgeting, performance indicators, etc.) and qualitative (expectations, procedures, etc.) aspects, it has to be able to address both semantic-logic and crisp numbers. Logical Multi-criteria decision-making (MCDM) methods are also preferable over other alternative decision-making frameworks because MCDM methods have "the potential capability of improving the *transparency*, *analytic rigor*, *auditability* and *conflict resolution* of decision-makers" [31]. Correspondingly, the MCDM provides:

- Means to establish accountability and transparency behind decisions, which may otherwise have unclear rationale and motives [25] by: placing stress on clearly stating and weighting the decision criteria, thereby improving transparency, and by ensuring that decisions taken through this method are explicit, paving the way to audit past decisions and thus provide accountability [32].

- Means for conflict resolution. This becomes a crucial issue when multiple perspectives are applied to a single smart critical infrastructure management decision [20, 24].

- Path for engagement and participation. Besides aiding decisions related to engineering, scientific studies, and cost analysis, one aspect that is becoming very crucial in decision-making studies is the engagement of multi-stakeholders and participation of communities [32].

The project considered various in-depth MCDM approaches that were used in other projects such as AIRM, PROMETHEE, and ELECTRE. However, during the eight case studies included in the SmartResilience and InfraStress projects, all of which involve end-user-owners of smart critical infrastructures, it became clear that the complexity of these methods made understanding them much more difficult and, at the same time, the required processing of the data needed proving to be prohibitively time consuming and expensive. Once an analysis is prepared and assessment data is input into the model (available on the project's ResilienceTool), the different optimization alternatives are scored following the combination of the user's input with the weighted criteria to rank the alternatives.

4. Implementation of the ResilienceCube concept in the "ResilienceTool" and merging it with the situational awareness systems

To support the methodology, a complete online tool was developed in which all the aspects described above were implemented with its intended user in mind - the person within a city or area, or a specific smart critical infrastructure [13]. The tool is based on the concept and its methodologies (the Cube, **Figure 11**), on the data resulting from extremely wide use (over 5,000 issues/indicators, over 300 assessments). In addition to the tools needed to support the ResilienceCube related analysis, presented above (database, methodologies, reporting), the tool contains also the Moodle-based education platform, support for standardization, a knowledge base (e.g. glossary) and a series of own and external tools linked to the system. Currently over a dozen of subsystems, containing all the features of the full system, but operating on the respective "private" databases are available for external users opted for the use of the system.

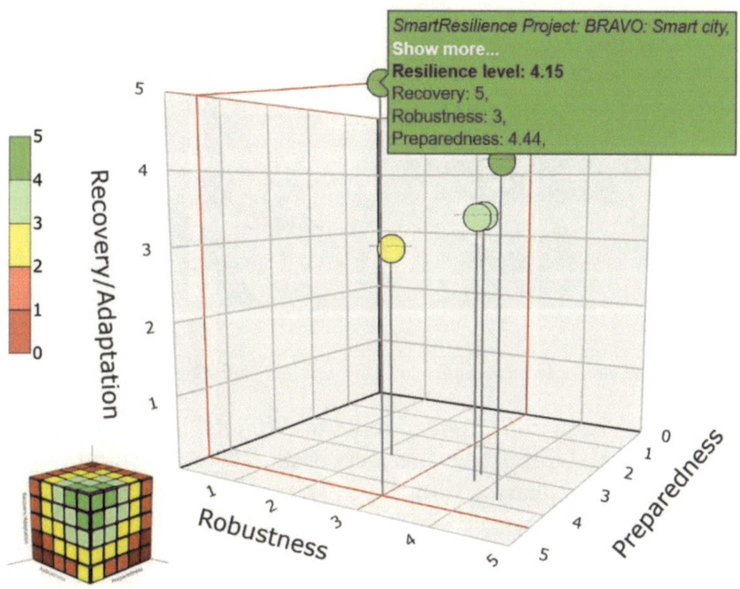

Figure 11.
ResilienceTool: The ResilienceCube.

5. Application of the concept and the tool

The project [1], covered over 30 case studies, (e.g. **Figures 8** and **12**).

6. Towards integration of resilience and situational awareness

Following the generally accepted position, that integration of all the aspects (concepts, data, tools, policies, implementation, etc.) is essential for successful risk and resilience governance, the InfraStress project of the EU [2] has developed an integrated framework (**Figure 13**).

Point	Real Time	Relative Time (h)	Description
	01.11.2017 10:00:00	0	Normal operational time
t0-t1			
t1	01.11.2017 11:00:00	1	Petrol and ship tanker commence the regular unloading operation
t1-t2			
t2	01.11.2017 14:00:00	4	A flying drone is briefly spotted just north of third basin area in Port of Koper
t2-t3			
t3	01.11.2017 14:00:30	4.0083	Petrol's personnel at the terminal briefly spot the approaching drone
t3-t4			
t4	01.11.2017 14:00:35	4.0097	Drone hits the pier's equipment, likely targeting the unloading arm close to the interface with the ship. Explosion.
t4-t5			
t5	01.11.2017 14:00:36	4.01	Equipment perforated, oil product (gasoline) massively released, fire.
t5-t6			
t6	01.11.2017 14:02:36	4.0433	Emergency declared - mobilization of the response teams
t6-t7			All activities in the second basin are stopped due to emergency situation
t7	01.11.2017 14:07:36	4.1267	On-shore response teams arrive to the scene – start of firefighting
t7-t8			
t8	01.11.2017 14:14:36	4.2433	Firefighting ships and maritime protection services arrive to the scene – start of firefighting
t8-t9			
t9	01.11.2017 14:34:36	4.5767	Firefighting successful
t9-t10			
t10	01.11.2017 20:34:36	10.5767	Area clean-up completion (oil products, smoke, foam, shore and sea)

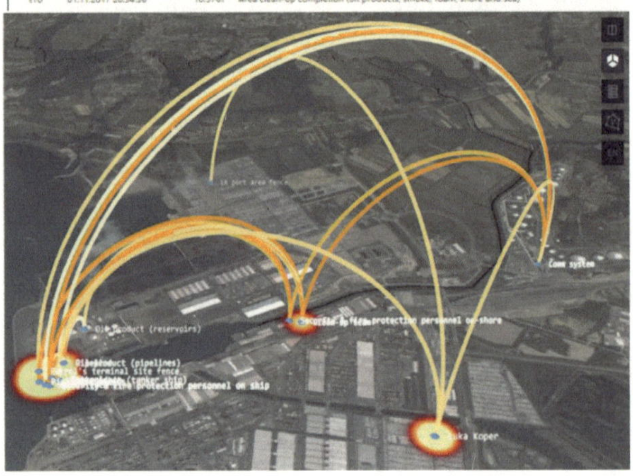

Figure 12.
Visualization of interdependencies based on indicators: User-oriented (InfraStress project).

The approach has been implemented in its five "pilots", cases covering "sensitive industrial plants and infrastructures", exposed to cyber-physical threats. The pilots cover chemical and pharmaceutical plants, ports, industrial zones, petrochemical plants, storage plants and similar. For all the plants the resilience has been analyzed, the analysis integrated with analysis of situational awareness systems performance (e.g. anti-drone systems or cyber protection systems), and, finally embedded into a testbed stress-testing concept for different scenarios.

7. Standardizing the concept: ISO 31050

The main calling of ISO 31050 (ISO New Work Item (NWI) 31050 "Guidance for Managing Emerging Risks to Enhance Resilience"[4]), is to provide universal, yet meaningful guidance on developing new competencies and business models to create relevant and realistic recommendations in an ever-changing uncertain world. The standard itself aims to provide the much-needed foresight and insight to deal with the rapidly changing landscape of risk due to the slew of new uncertainties and new emerging risks, the management of which is essential for society. It is based on the idea that these, emerging risks, are those that can challenge the resilience of the critical infrastructures the most. It aims to integrate and align the (emerging) risk

4 https://www.iso.org/standard/54224.html

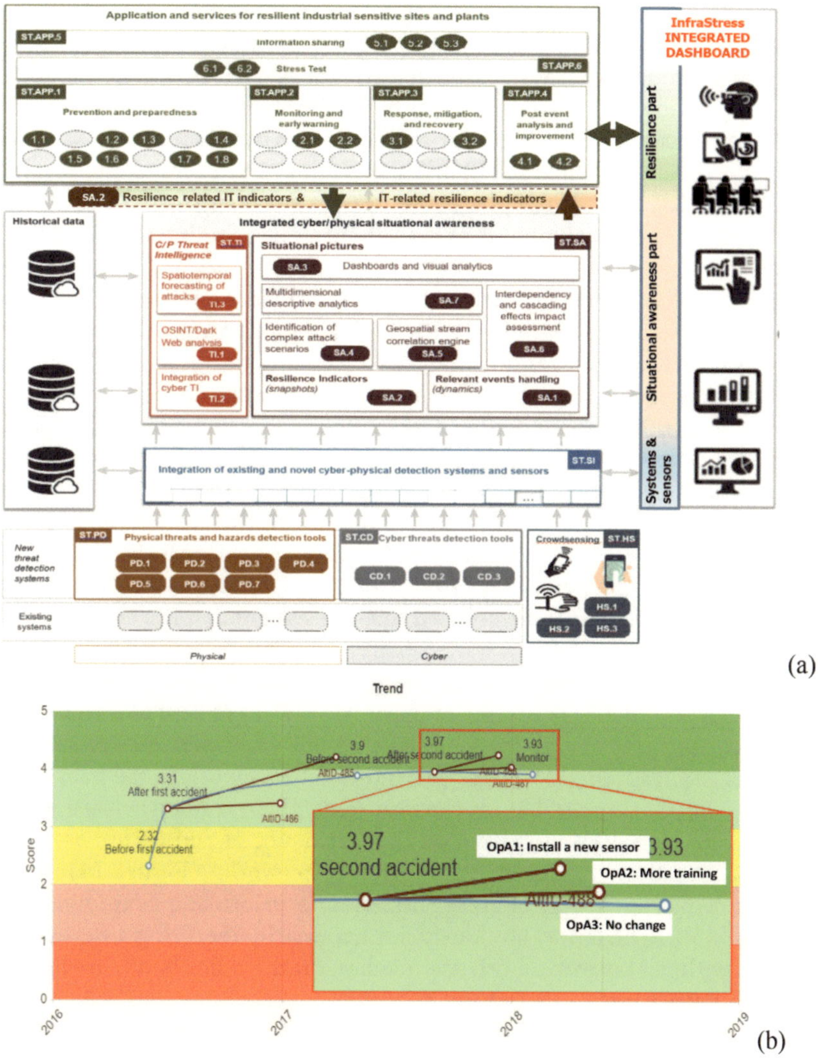

Figure 13.
InfraStress framework integrating resilience analysis and situational awareness and its application to resilience improvement decision-making: Within the overall framework (a), the embedded MCDM modules communicate with other modules and get values through a Kafka broker, and lead to the resilience assessment based decision optimization (b).

framework with resilience framework (definitions, concepts, requirements) and propose outputs such as a procedure for scanning for emerging risks, metrics for assessing possible impacts of those risks on critical infrastructure's resilience. The management framework, guidance for interoperability and common/agreed indicators, as well as the particular considerations related to emerging risks in resilience assessment. ISO 31050 will be part of the ISO 31000:2018 family of standards, monitored by the ISO Technical Committee TC262.

8. Conclusions

The ResilienceCube allows presenting the resilience of a critical infrastructure as a single point (Resilience Index) in a 3D space. The concept, especially as implemented in the tool (the ResilienceTool) is user-friendly, intuitively understandable and flexible. It supports end-users (authorities, critical infrastructure operators and owners) in improving the disaster resilience of respective critical infrastructures through indicator-based assessment of their resilience capabilities. This solution provided by SmartResilience and InfraStress projects is oriented towards the practical needs of end-users and has been developed in close collaboration with all relevant stakeholders. In order to achieve the Technology Readiness Level (TRL) beyond the initially planned 4, the Tool is being tested and constantly improved through the development of realistic use cases, both within and beyond the projects.

The SmartResilience and InfraStress ResilienceTool are envisaged to stay available, free of charge for the registered ERRA members, also after the project end. The main ERRA service (risk and resilience "Assessment-as-a-service") will be performed by the Agency together with and subcontracting to Agency member organizations (organizational members and individuals) which have the different competencies needed to meet the specific needs of specific industry branches or application areas (e.g. critical infrastructures or new technologies). In the most general terms, ERRA would contact and negotiate with the customers, engage the experts among the Agency members, process the contracts with the customer, and guarantee the quality of assessment provided by the Agency. Main Agency services would be the self-assessment, the audited self-assessment and the third party audit, similarly to the services of GRI (www.globalreporting.org).

The concepts and the tools were applied to the analysis of health infrastructures (over 100 hospitals) in a COVID-like scenario [33]. The concept allows integrating the qualitative approaches with those based on a more complex quantitative resilience analysis (e.g. [30, 34, 35] or [22]). In addition, the work in the background of this paper has clearly shown, that the current research on resilience has a number of different aspects: from those focusing on the "resilience of and within a network" (e.g. in the area of electric grids or transportation networks - **Figure 14**), to those looking at resilience as "ability of an organization to absorb and adapt in a changing environment" [36]. The latter, obviously not necessarily requiring a network, or measuring it within a network. Both approaches, on the other hand, are applicable to critical infrastructures.

To conclude, within the plethora of the "current" existing tools (e.g. those presented or reviewed in [25, 37–42] or [43]), that all can simulate different resilience aspects of large and complex systems and/or apply optimization techniques to improve it (e.g. by indicating the optimum path towards system recovery or improving preparedness to unknowns) the approach presented here proposes a pragmatic and flexible way to achieve improvement through applying resilience indicators. It has been "combat-tested" in a number of large-scale cases and it has confirmed being robust and combinable with the systems previously on site.

Finally, the concepts might have one of an even more ambitious potential allocation: the biggest infrastructure of all is the "infrastructure of all infrastructures" of our planet Earth and the "global society". Technically, the methodology presented here can be applied for this case too, allowing to quantify the global

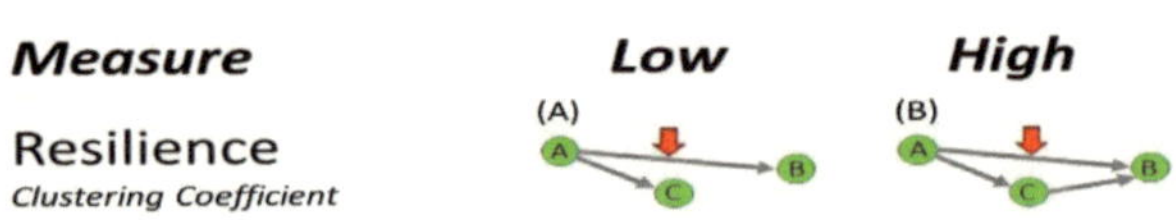

Figure 14.
Resilience of a network (graph representation) – Not always the same as the engineering resilience of an organization, defined by ISO as "ability of an organization to absorb and adapt in a changing environment" (ISO 23316, https://www.iso.org/obp/ui#iso:std:iso:22316:ed-1:v1:en).

resilience (note: we do not have anything better around yet!) and point out where the "investment in the improvement of the global infrastructure" will be the most effective and beneficial.

Acknowledgements

The contribution is based on the Grant Agreements No. 700621 and No. 833088 supporting the work on the SmartResilience and InfraStress projects, respectively, provided by the Research Executive Agency (REA) ('the Agency'), under the power delegated by the European Commission ('the Commission'). This support is gladly acknowledged. The work presented is a result of a great collective effort of all the partners of the SmartResilience and InfraStress projects – many thanks for this to all of them. Nevertheless, some of these contributions were particularly important for this particular paper, e.g., the methodological ones of K. Øien, big-data related ones of P. Klimek, and technical preparatory ones of Sh. Macika, M. Th. Nguyen and T. Rosen. The authors mention this with appreciation and gratitude.

Appendix 1

Example of the resilience indicator data sheet in SmartResilience and InfraStress systems (a) and their implementation in the database (b).

(a)

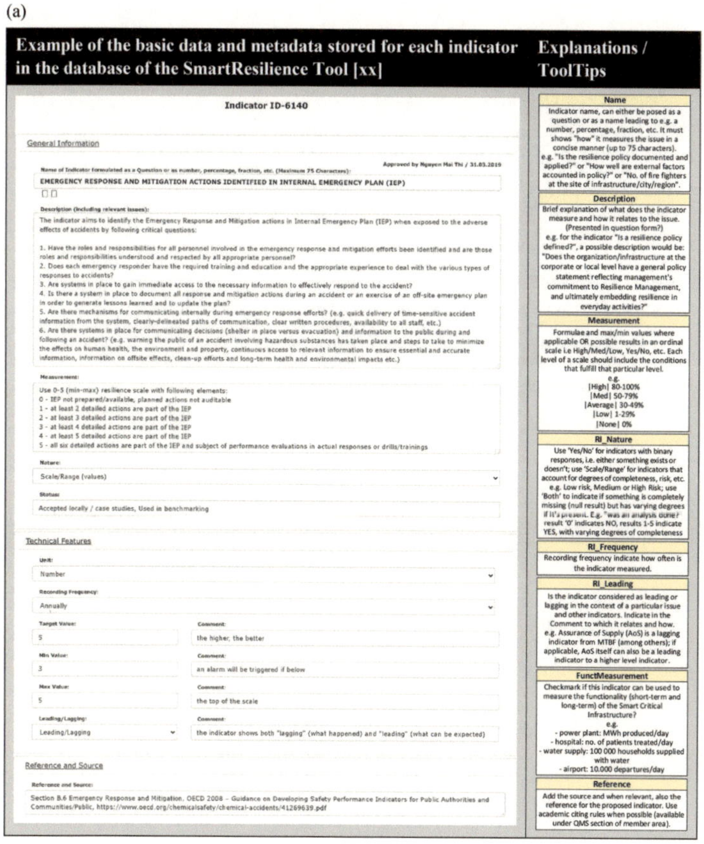

(b)

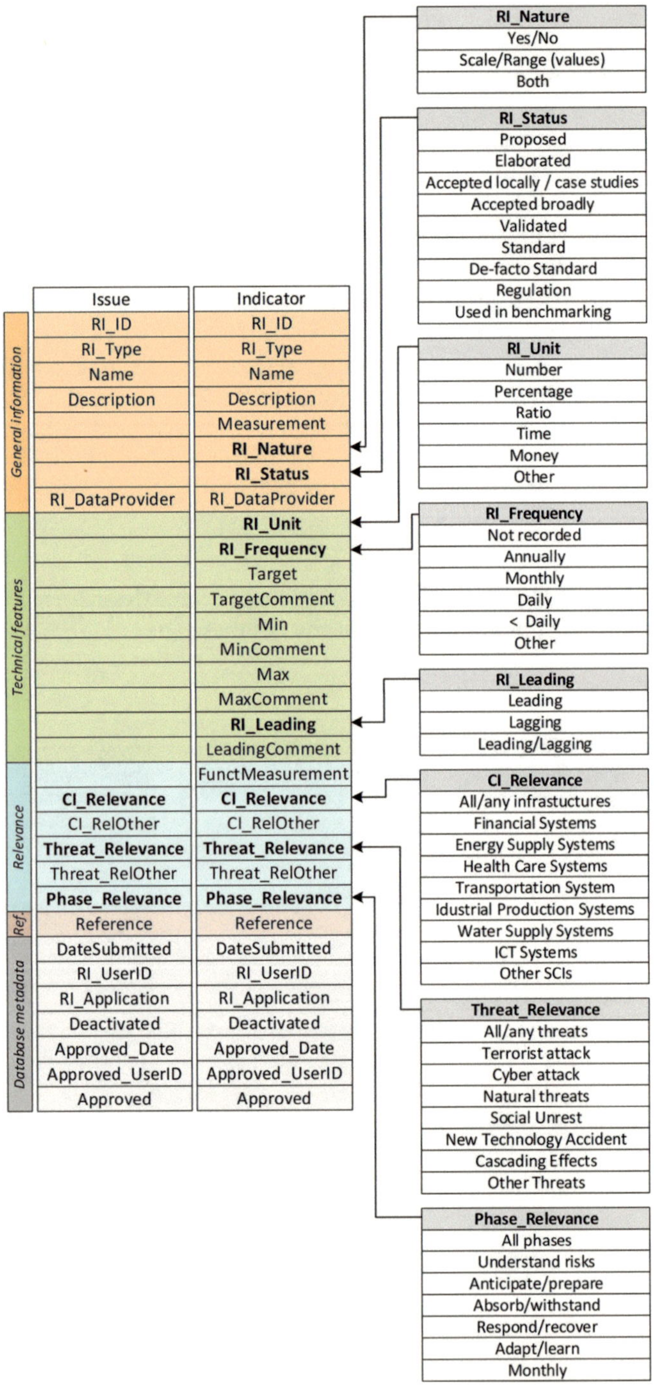
Issue
Indicator
General information
Technical features
Relevance
Ref
Database metadata
RI_ID
RI_Type
Name
Description
Measurement
RI_Nature
RI_Status
RI_DataProvider
RI_Unit
RI_Frequency
Target
TargetComment
Min
MinComment
Max
MaxComment
RI_Leading
LeadingComment
FunctMeasurement
CI_Relevance
CI_RelOther
Threat_Relevance
Threat_RelOther
Phase_Relevance
Reference
DateSubmitted
RI_UserID
RI_Application
Deactivated
Approved_Date
Approved_UserID
Approved
RI_Nature
Yes/No
Scale/Range (values)
Both
RI_Status
Proposed
Elaborated
Accepted locally / case studies
Accepted broadly
Validated
Standard
De-facto Standard
Regulation
Used in benchmarking
RI_Unit
Number
Percentage
Ratio
Time
Money
Other
RI_Frequency
Not recorded
Annually
Monthly
Daily
< Daily
Other
RI_Leading
Leading
Lagging
Leading/Lagging
CI_Relevance
All/any infrastructures
Financial Systems
Energy Supply Systems
Health Care Systems
Transportation System
Idustrial Production Systems
Water Supply Systems
ICT Systems
Other SCIs
Threat_Relevance
All/any threats
Terrorist attack
Cyber attack
Natural threats
Social Unrest
New Technology Accident
Cascading Effects
Other Threats
Phase_Relevance
All phases
Understand risks
Anticipate/prepare
Absorb/withstand
Respond/recover
Adapt/learn
Monthly

Author details

Aleksandar S. Jovanovic[1,2]*, Somik Chakravarty[1,2] and Marjan Jelic[2]

1 European Risk and Resilience Institute (EU-VRi), Stuttgart, Germany

2 Steinbeis Advanced Risk Technologies (R-Tech), Stuttgart, Germany

*Address all correspondence to: jovanovic@eu-vri.eu;
jovanovic@risk-technologies.com

IntechOpen

© 2021 The Author(s). Licensee IntechOpen. This chapter is distributed under the terms of the Creative Commons Attribution License (http://creativecommons.org/licenses/by/3.0), which permits unrestricted use, distribution, and reproduction in any medium, provided the original work is properly cited.

References

[1] SmartResilience (2016). Smart Resilience Indicators for Smart Critical Infrastructures – The European Union's Horizon 2020 Research and Innovation Programme, Grant Agreement No 700621 (2016-2019). Coordinator: EU-VRi, www.smartresilience.eu-vri.eu.

[2] InfraStress: Improving resilience of sensitive industrial plants & infrastructures exposed to cyber-physical threats, by means of an open testbed stress-testing system; Project reference: 833088, Project type: Innovation Action, Call: SU-INFRA01-2018-2019-2020 - Prevention, detection, response and mitigation of combined physical and cyber threats to critical infrastructure in Europe https://www.infrastress.eu

[3] European Commission (2013). DRS-14-2015: Topic: Critical Infrastructure Protection topic 3: Critical Infrastructure resilience indicator – analysis and development of methods for assessing resilience, https://ec.europa.eu/research/participants/portal/desktop/en/opportunities/h2020/topics/143-drs-14-2015.html

[4] Jovanović, A., Bellini, E. (Eds.), Aligning the resilience-related research efforts in the EU-DRS projects, Joint Workshop DRS-7&14 projects | Brussels, September 13–14, 2017, Steinbeis Edition, Stuttgart, Germany 2017, ISBN 978-3-95663-143-6 2017 |E-Book (PDF), 165 p.

[5] Brown, B., Neil-Adger, W., Tompkins, E., Bacon, P., Shim, D. and Young, K. (2001). Trade-off analysis for marine protected area management, Ecological Economics, vol. 37, no. 3, pp. 417–434.

[6] Jovanović, A., Klimek, P., Choudhary, A., Schmid, N., Linkov, I., Øien, K., ... Lieberz, D. (2016). SmartResilience D1.2: Analysis of existing assessment resilience approaches, indicators and data sources: Usability and limitations of existing indicators for assessing, predicting and monitoring critical infrastructure resilience, EU project SmartResilience https://www.smartresilience.eu-vri.eu , Project No. 700621 (2016-2019), Contact: EU-VRi, Stuttgart, Germany.

[7] IEC 61508 (2010). Functional Safety of Electrical/Electronic/Programmable Electronic Safety-Related Systems. Part 1-7. Geneva: International Electrotechnical Commission

[8] SmartResilience (2017). Deliverable D 3.2: Assessing resilience of smart critical infrastructures based on indicators http://www.smartresilience.eu-vri.eu/sites/default/files/publications/SmartResD3.2.pdf.

[9] SmartResilience (2017). Deliverable D 5.1: Report on the results of the interactive workshop http://www.smartresilience.eu-vri.eu/sites/default/files/publications/SmartResD5.1.pdf

[10] SmartResilience (2017). Deliverable D2.1: Understanding "smart" technologies and their role in ensuring resilience of infrastructures, EU project SmartResilience, Project No. 700621 (2016-2019), Contact: EU-VRi, Stuttgart, Germany

[11] SmartResilience (2017). Deliverable D3.1: Contextual factors related to resilience, EU project SmartResilience, Project No. 700621 (2016-2019), Contact: EU-VRi, Stuttgart, Germany http://www.smartresilience.eu-vri.eu/sites/default/files/publications/Smart ResD3.1.pdf

[12] SmartResilience (2017). Stress test and evaluation framework. EU project SmartResilience, Project No. 700621 (2016-2019), Contact: EU-VRi, Stuttgart, Germany. http://smartresilie

nce.eu-vri.eu/sites/default/files/publications/SmartResD5.2.pdf

[13] Jovanovic, A., M. Jelic, T. Rosen, P. Klimek, S. Macika, and K. Øien (2019). SmartResilience D3.7: "The ResilienceTool" of the Smart-Resilience project. EU project SmartResilience, Project No. 700621. http://www.smartresilience.eu-vri.eu/sites/default/files/publications/SmartResD3.7.pdf

[14] Fisher, R.E., Bassett, G.W., Buehring, W.A., Collins, M.J., Dickinson, D.C., Eaton, L.K., ... Peerenboom, J.P. (2010). Constructing a Resilience Index for the Enhanced Critical Infrastructure Protection Program, Argonne National Laboratory, Decision and Information Sciences Division, ANL/DIS-10-9, Argonne, IL, USA http://www.ipd.anl.gov/anlpubs/2010/09/67823.pdf

[15] EPRI (2000). Guidelines for Trial Use of Leading Indicators of Human Performance: The Human Performance Assistance Package. EPRI (U.S. Electric Power Research Institute), Palo Alto, CA, 10000647.

[16] EPRI (2001). Final report on Leading Indicators of Human Performance. EPRI, Palo Alto, CA, and the U.S. Department of Energy, Washington, DC, 1003033.

[17] Øien, K. (2001). A framework for the establishment of organizational risk indicators. Reliability Engineering and System Safety, 74, 147–167.

[18] National Infrastructure Advisory Council. (2009) Critical Infrastructure Resilience, Final Report and Recommendations, U.S. Department of Homeland Security, Washington, D.C., available at http://www.dhs.gov/xlibrary/assets/niac/niac_critical_infrastructure_resilience.pdf

[19] Øien, K., Massaiu, S., & Tinmannsvik, R.K. (2012). Guideline for implementing the REWI method; Resilience based Early Warning Indicators. SINTEF report A22026, Trondheim, Norway.

[20] Cai, X., Lasdon, L. and Michelsen, A.M. (2004). Group decision-making in water resources planning using multiple objective analysis, Journal of Water Resources Planning and Management, vol. 130, no. 1, pp. 4–14.

[21] Øien, K., A. Jovanović, et al. (2018). D3.6 Guideline for assessing, predicting and monitoring resilience of Smart Critical Infrastructures, , EU project SmartResilience https://www.smartresilience.eu-vri.eu , Project No. 700621 (2016-2019), Contact: EU-VRi, Stuttgart, Germany.

[22] Tofani, A., D'Agostino, G., Di Pietro, A., Rosato, V. (2019). Modeling Resilience in Electrical Distribution Networks. In book: Management of Critical Infrastructure, November 2019 (10.5772/intechopen.85917)

[23] Zobel, C.W. (2014). Quantitatively representing nonlinear disaster recovery. *Decision Sciences*, 43(4), 687-710.

[24] Mustajoki, J., Hamalainen, R.P. and Marttunen, M. (2004). Participatory multicriteria decision analysis with Web-HIPRE: A case of lake regulation policy. Environmental Modeling & Software, vol. 19, no. 6, pp. 537–547.

[25] Boumphrey, R., and Bruno, M. (2015). "Foresight Review of Resilience Engineering: designing for the expected and unexpected."

[26] Business Dictionary (2017). Downtime. Business Dictionary. http://www.businessdictionary.com/definition/downtime.html

[27] Sahebjamnia, N., Torabi, S. A., Mansouri, S. A. (2015). Integrated business continuity and disaster

recovery planning: Towards organizational resilience. European Journal of Operational Research, 242(1), 261-273. https://doi.org/10.1016/j.ejor.2014.09.055

[28] Ruiying. Li. et.al. (2017). A new resilience measure for supply chain networks. MDPI. Sustainability. 9(1), 144. doi:10.3390/su9010144

[29] Zhao. S., Liu. X., Zhuo. Y. (2017). Hybrid Hidden Markov Models for resilience metrics in a dynamic infrastructure system. Reliability engineering and safety systems. Elsevier. http://www.sciencedirect.com/science/article/pii/S095183201730248X accessed on November 17, 2017

[30] Ganin et al (2017). Resilience and efficiency in transportation networks. Science Advances, 3(12): e1701079

[31] Kabir, G., Sadiq, R. and Tesfamariam, S. (2014). A review of multi-criteria decision-making methods for infrastructure management, Structure and Infrastructure Engineering, vol. 10, no. 9, pp. 1176-1210.

[32] Greiner, R., Herr, A., Brodie, J., and Haynes, D. (2005). A multi-criteria approach to great barrier reef catchment (Queensland, Australia) diffuse-source pollution problem, Marine Pollution Bulletin, vol. 51, no. (1–4), pp. 128–137.

[33] Jovanović, A. et al. (2020): Assessing resilience of healthcare infrastructure exposed to COVID-19: emerging risks, resilience indicators, interdependencies and international standards; Environment Systems and Decisions, June 2020 https://doi.org/10.1007/s10669-020-09779-8

[34] Fox-Lent, C., & Linkov, I. (2018). Resilience Matrix for Comprehensive Urban Resilience Planning. In: Resilience-Oriented Urban Planning (pp. 29-47). Springer, Cham

[35] Fox-Lent, C., Bates, M.E., & Linkov, I. (2015). A matrix approach to community resilience assessment: an illustrative case at Rockaway Peninsula. Environment, Systems, and Decisions, 35: 209-218

[36] ISO23316 - Security and resilience — Organizational resilience — Principles and attributes, ISO https://www.iso.org/standard/50053.html

[37] Ghafir, I., Saleem, J., Hammoudeh, M., Faour, H., Prenosil, V., Jaf, S., Jabbar, S., and Baker, T. (2018). "Security threats to critical infrastructure: the human factor." *The Journal of Supercomputing*, 74(10), 4986– 5002.

[38] Gouglidis, A., Shirazi, S. N., Simpson, S., Smith, P., and Hutchison, D. (2016). "A multi-level approach to resil-ience of critical infrastructures and services." *2016 23rd International Conference on Telecommunica- tions (ICT)*, IEEE, Thessaloniki, Greece, 1–5.

[39] König, S., Schaberreiter, T., Rass, S., and Schauer, S. (2019). "A Measure for Resilience of Critical Infrastruc- tures." *Critical Information Infrastructures Security*, E. Luiijf, I. Žutautaitė, and B. M. Hämmerli, eds., Springer International Publishing, Cham, 57–71.

[40] Martin, H., and Ludek, L. (2013). "The status and importance of robustness in the process of critical infrastructure resilience evaluation." *2013 IEEE International Conference on Technologies for Homeland Secu- rity (HST)*, IEEE, Waltham, MA, USA, 589–594.

[41] Royal Academy of Engineering. (2018). *Cyber safety and resilience - strengthening the digital systems that support the modern economy*. London.

[42] Tokgoz, B. E., and Gheorghe, A. V. (2013). "Resilience quantification and its application to a residential build- ing

subject to hurricane winds." *International Journal of Disaster Risk Science*, 4(3), 105–114.

[43] Walker-Roberts, S., Hammoudeh, M., and Dehghantanha, A. (2018). "A Systematic Review of the Availability and Efficacy of Countermeasures to Internal Threats in Healthcare Critical Infrastructure." *IEEE Ac- cess*, 6, 25167–25177.

Section 2

Dependency Mechanisms and Systems Survivability

Chapter 5

Integrating Resilience in Time-based Dependency Analysis: A Large-Scale Case Study for Urban Critical Infrastructures

Vittorio Rosato, Antonio Di Pietro, Panayiotis Kotzanikolaou, George Stergiopoulos and Giulio Smedile

Abstract

As critical systems shall withstand different types of perturbations affecting their functionalities and their service level, resilience is a very important requirement. Especially in an urban critical infrastructures where the occurrence of natural events may influence the state of other dependent infrastructures from various different sectors, the overall resilience of such infrastructures against large scale failures is even more important. When a perturbation occurs in a system, the quality (level) of the service provided by the affected system will be reduced and a recovery phase will be triggered to restore the system to its normal operation level. According to the implemented recovery controls, the restoration phase may follow a different growth model. This paper extends a previous time-based dependency risk analysis methodology by integrating and assessing the effect of recovery controls. The main goal is to dynamically assess the evolution of recovery over time, in order to identify how the expected recovery plans will eventually affect the overall risk of the critical paths. The proposed recovery-aware time-based dependency analysis methodology was integrated into the CIPCast Decision Support System that enables risk forecast due to natural events to identify vulnerable and disrupted assets (e.g., electric substations, telecommunication components) and measure the expected risk paths. Thus, CIPCast can be valuable to Critical Infrastructure Operators and other Emergency Managers involved in a crisis assessment to evaluate the effect of natural and anthropic threats affecting critical assets and plan proper countermeasures to reduce the overall risk of degradation of services. The proposed methodology is evaluated in a real scenario, which utilizes several infrastructures and Points of Interest of the city of Rome.

Keywords: time, resilience, dependency, critical infrastructure, impact, energy, urban, telecommunications, graph, chain, cascading, risk management, risk analysis

1. Introduction

Critical infrastructures consist of physical and cyber assets, systems, and networks, that are essential for the functioning of a society and economy. The damage

IntechOpen

to a critical infrastructure, caused by natural (e.g., earthquakes, fire) or anthropic (e.g., hacking, sabotage, vandalism) events may produce a significant negative impact for other systems and thus amplify the effects and reducing the system capability to return to an equilibrium state.

In a scenario consisting of multiple infrastructures with several dependencies among them, the implementation of mitigation controls that may affect the resilience level of the systems, is valuable to preserve and restore the essential societal services. Since resilience-related controls will positively affect the capability of a system to resist, absorb, adapt and/or recover from the effects of a hazard in a timely and efficient manner, it is important to analyse the effect of such controls, in order to support decision making related to the selection and prioritization of alternative mitigation controls. For example, when electric transmission or distribution networks are affected by disturbances such as floods, in general, mitigation and restoration actions are performed through protection and automation devices and manual interventions to reduce the duration of the outage and preserve the power supply to critical systems such as hospitals [1–3].

In the US, in order to support the different players involved in modeling, simulation, and analysis of the nation's critical infrastructures, the National Infrastructure Simulation and Analysis Center (NISAC) was established. NISAC analysts assess critical infrastructure risk, vulnerability, interdependencies, and event consequences. In Europe, in order to support the different players involved in the resilience enhancement, emergency and response management of critical infrastructures to natural and man-made hazards, the Infrastructure Simulation and Analysis Centre (EISAC) is aiming at establishing a collaborative, European-wide network of national centres empowered by core technologies.

This paper extends a recent work on critical infrastructure dependency analysis and introduces time-based analysis models to study the evolution of restoration actions in a scenario of dependent systems. This model was integrated into CIPCast Decision Support System, named CIPCast hereafter, that is part of the on-going products and activities developed in the context of the Italian node of EISAC, called I-EISAC, aiming to support infrastructure and civil protection operators operators in the risk assessment of critical infrastructures.

CIPCast can provide an operational (24/7) forecast and risk analysis for different infrastructures in a specific area showing risk maps of infrastructure elements which could be damaged by different events e.g. earthquakes. In particular, CIPCast allows: (i) Assessing the seismic vulnerability of different EDNs components; (ii) estimating possible earthquake-induced physical damage; (iii) estimating the impact on service(s) functionality in terms of outage duration associated with the predicted physical damage and considering the known inter-dependencies; (iv) estimating the consequences of the predicted outages, according to several metrics accounting for economic losses and reduction of citizens well-being.

The remainder of the paper is organized as follows. Section 2 presents related works in the area. In Section 3, we introduce notions of time-based and resilience-aware dependency analysis. In Section 4, we apply the analysis to a case study related to the area of Rome. Finally, in Section 5, some conclusions and ideas for future works are drawn.

2. Related work

Modeling critical infrastructures and urban systems for risk assessment purposes is a well-known and established research field. Preliminary work that laid the foundation in this area is often attributed to Rinaldi et al., first in [4] where authors categorised dependencies in critical infrastructures as Physical, Cyber/informational, Geographic, Logical and Social dependencies, and later in where authors created taxonomies for disruptions or outages and marked them as cascading, escalating, or common-cause [5]). Critical infrastructure modeling events where first defined as cascade initiating (i.e., an event that causes an event in another CI) and cascade resulting (i.e., an event that results from an event in another CI) by the empirical study of Van Eeten et al. [6].

Basic modeling approaches usually fall within one of the following six categories categories [5, 7]:

1. Aggregate supply and demand tools, which evaluate the total demand for infrastructure services in a region and the ability to supply those services

2. Dynamic simulations, which analyze the effects of disruptions, and their associated consequences.

3. Agent-based models, which model operational attributes and states of infrastructure operation; usually on a graph model.

4. Physics-based models, which utilize standard engineering techniques such as power flow and stability analyses for electric power grids.

5. Population mobility models that focus on geospatial movement.

6. Leontief input–output models, which utilize linear, time-independent analysis of commodities among infrastructure sectors.

Our approach can be classified as both dynamic simulation and agent-based model. It utilizes operational attributes to model interdependencies in urban environments as a graph, while still allowing for dynamic input of data in order to analyze the effects of disruptions in the urban web along with quantifying their associated consequences.

Each critical infrastructures sector has its own group of research publications that utilize some of the aforementioned techniques to model and analyse risk. For example, in the water sector, OpenMI [8] supports federated modeling and simulation for water systems, while multiple publications exist that analyze interdependencies at the transportation sector using traffic flow simulation models [9], Bayesian networks to model the correlation structure of highway networks [10] etc. The Energy sector is also a highly researched area. Wide Area Measurement Systems (WAMS) have been extensively researched, especially for the detection of optimal locations for metering device placement, in order to achieve increased robustness of the WAMS infrastructure. Modeling and quantifying dependencies between the electrical and information infrastructures of WAMS in smart grids has been recently studied in [11]. Topological observability of power systems has been fully described in [12]. Still, cross-sector approaches do exist that opt to combine combine models from multiple sectors and enable integrated or federated simulations. Some examples include DIESIS [13] and EPIC [14].

The North American Electric Reliability Corporation (NERC) has recently developed Critical Infrastructure Protection (CIP) standards which introduce cyber security compliance requirements for power systems [15]. Various research has developed methodologies that aim to quantify these requirements. In [15], authors proposed a risk-based dependency analysis for modeling and quantifying dependencies over time, which was also later used in [11] along with electrical centrality metrics to quantify the level of each dependencies in the smart grid. A different approach for simulating common-cause and cascading effects was also introduced by the authors in [16]. Similarly, authors in [17] proposed to use access graph models to analyze trust between systems and the security exposure of a large scale smart grid environments. In [18], authors developed a graph-based workflow model for assessing the security risks from cybersecurity incidents on electric grids and build relevant scenarios.

The presented approach is mostly based on the methodologies presented in [15]. We aggregate data into dependency matrices and utilize models from real-world urban systems to map them into dependency graphs. The presented approach is based on network modeling and path analysis. It depicts dependencies of the connected urban infrastructures as a graph and identifies high risk, critical paths that are either modeled as flows of information, power or other related type of dependency. Similar techniques have been used in uniform [19, 20] or flow models [12, 21].

3. Time-based and resilience-aware dependency analysis

3.1 Definitions and set up

We consider a directed graph $G = (V, E)$ where $V = \{v_i\}, i = 1, \dots m$, is the set of nodes (infrastructures, components or Point of Interest–POIs hereafter) and $E = \{e_{ij}\}$ is the set of edges (or dependencies) and $deg(v_i)$ is the degree of node v_i. An edge e_{ij} from node v_i to v_j denotes a dependency (and consequently a risk relation) denoted with $v_i \to v_j$ that is derived from the dependence of node v_j on a service provided by node v_i. A dependency is defined as a "one-directional reliance of an asset, system, network or collection thereof – within or across sectors – on an input, interaction or other requirement from other sources in order to function properly" [22]. A node could thus represent a *consumer* or a *producer* of a service provided by another node (or both), depending on its role in the system.

Our model extends the cumulative dependency risk model of [23, 24]. Without loss of generality, let $v_0 \to v_1 \to \dots \to v_n$ be a dependency chain, involving $n + 1$ nodes and their corresponding n dependencies. Let L_{v_{j-1}, v_j} be the likelihood that a disruptive event (threat) that happened in node v_{j-1} will also affect (cascade) to node v_j due to their dependency and let I_{v_{j-1}, v_j} be the relevant impact (damage) caused to v_j. We should note here that L is not the likelihood of threat manifestation, but rather the likelihood of an already manifested threat to cascade (i.e. affect) different nodes.

Based on the definitions of [23], the risk exhibited by a node due to its n-th order dependency is defined as:

$$R_{v_0, \dots, v_n} = L_{v_0, \dots, v_n} \cdot I_{v_{n-1}, v_n} \equiv \prod_{i=0}^{n-1} L_{v_i, v_{i+1}} \cdot I_{v_{n-1}, v_n}. \tag{1}$$

Then the *cumulative dependency risk* which includes the *overall* risk exhibited by all the nodes within the sub-chains of an n-order dependency is defined as:

$$DR_{v_0, \dots, v_n} = \sum_{i=1}^{n} R_{v_0, \dots, v_i} \equiv \sum_{i=1}^{n} \left(\prod_{j=1}^{i} L_{v_{j-1}, v_j} \right) \cdot I_{v_{i-1}, v_i}. \tag{2}$$

3.2 Extending the model for resilience

Let $\mathbb{T} = \{threat\}$ be the set of k natural or human-related threats that may affect the quality of service provided by the generic node v_i. The damage $D_i(t)$ associated with the perturbation t is usually an s-shaped function. Let $\mathbb{C}^{v_i} = \{c_1^{v_i}, \dots, c_l^{v_i}\}$ be the set of l^{v_i} security controls that may be implemented in a system/infrastructure v_i to improve their resilience against threats (e.g. restoration security controls, redundancy security controls etc).

By combining Resilience and Threat variables with the directed graph model of interdependent POIs, we can perform a granular analysis of the risk imposed by POI interdependencies based on their risk and resilience levels. We opt to use the multi-risk dependency analysis method as proposed in [23–25] and implemented later in [15].

3.3 Resilience mapping

A many-to-many mapping may exist between the threats and the security controls, i.e. a security control may mitigate, at some extent, one or more threats, while a security threat may require one or security controls. For each security control, different weights can be used to define the effectiveness of a control against different threats and also for their application to specific infrastructures. This is a realistic modeling of resilience, since many controls do not have the same effect against all threats and different infrastructures are benefited more than others from specific security controls, given the nature of the infrastructure and the intrinsic characteristics of each threat.

For example, if infrastructure (node) v_1 is affected by a power outage (i.e. the initiating threat event), then a node v_2 which is depended on v_1 might suffer a partial unavailability (modeled as impact I_{v_1,v_2}) at a certain extend quantified as the likelihood L_{v_1,v_2}. L_{v_1,v_2} depicts the possibility that a power outage would affect node v_2 and I_{v_1,v_2} depicts the amount of damage done to v_2 due to its partial unavailability incident.

In the aforementioned example, node v_1 could have implemented the use of a redundant power generator as a security control with quantified measurements (i) $\overline{L}_{v_1,v_2}$ and (ii) $\overline{I}_{v_1,v_2}$ depicting (i) the resilience influence of control c on node v_2 for the given threat (in our case, the power outage), and (ii) the extent of reduction to the initial estimated damage I_{v_1,v_2}, respectively. The existence of the control c will reduce the possibility of a power outage to affect v_2 by $\overline{L}_{v_1,v_2}$ percent, and/or the corresponding impact from the same threat on v_2 by $\overline{I}_{v_1,v_2}$.

Generalising this to n nodes, this gives us with a Resilience series calculation that can be depicted as follows:

$$Res_{v_0,\dots,v_n} = \sum_{i=1}^{n}\left(\prod_{j=1}^{i}\overline{L}_{v_{j-1},v_j}\right)\cdot\overline{I}_{v_{i-1},v_i} \tag{3}$$

where *Res* depicts the overall resilience of a network against a specific $threat \in \mathbb{T}$ when the security control c is implemented in all nodes. It should be noted, that the resilience expressed by Eq. (3) depicts the resilience of a network due to the existence and the efficacy of security control c. However, the Resilience of a network depends also on the vulnerability of the node v_j to specific threats that may produce a disservice of the network.

For example, if we consider an electric substation, in order to increase its resilience against a seismic threat, there might be several options aiming to reduce the likelihood of the threat that produces a failure and/or to reduce the magnitude of the impact e.g. to enhance the structural properties of the building or increment the number of technical crews so that in case of a failure the duration of outage can be reduced.

In a complex study of a large CI system, such as the city of Rome, the interplay among network topology, size, quality and distribution of technical systems along the network, emergency management ability do have an impact on the evolution and the duration of a crises and thus influence the system resilience. They have been thus studied in order to establish the "sensitivity" of the resilience score with respect to each one of the described properties [3].

Conveniently, the Resilience introduced by a security control against a specific threat on the entire network of interdependent nodes can be algorithmically modeled as a matrix multiplication. For the first matrix, columns represent existing nodes, while rows represent different security controls. Cell values depict the possibility of a security control to mitigate some part of the impact of a specific threat for each node present in the graph. The second matrix depicts the impact reduction that can be achieved by security controls onto the existing interdependent nodes. Similarly, columns represent existing nodes, while rows represent different security controls, but, here cell values depict the maximum potential impact reduction achieved at each node by the implementation of each security control. Thus, in this matrix, cells have negative values. Resilience is then modeled as the matrix multiplication of the two matrices (threat reduction and impact reduction matrices), as depicted in **Figure 1**.

3.4 Calculating cumulative dependency risk in the presence of resilience controls

By combining Eq. 1 and Eq. 2 with Eq. 3, the cumulative dependency risk in the presence of resilience controls can be defined as follows:

$$DR^{Res}_{v_1,\dots,v_n} = \sum_{i=1}^{n} \left[\left(\prod_{j=1}^{i} L_{v_{j-1},v_j} \right) \cdot I_{v_{i-1},v_i} - \left(\prod_{j=1}^{i} \overline{L}_{v_{j-1},v_j} \right) \cdot \overline{I}_{v_{j-1},v_j} \right] \tag{4}$$

As discussed above, $\overline{L}_{v_{j-1},v_j}$ introduces a likelihood for the security controls (actions). Specifically, it quantifies the possibility of one security control to mitigate some part of the impact of a threat.

Impact I in Eqs. 1 through 4 is assigned values that reflect the maximum expected impact for each modeled dependency. This first implies that eqs will always calculate produce the worst case cascading risk $DR^{Res}_{v_1,\dots,v_n}$, and also that all modeled dependencies exhibit the same impact growth rate; something that is not true in real-world situations, where different infrastructure resilience allows for different impact growth rates over time. Thus, we use the same modeling approach as in [15] and incorporate a dynamic time-based analysis model where $T_{i,j}$ denotes

Likelihood matrix for Threat *(threat)*				
Nodes / Controls	v_1	v_2	…	v_n
c_1	0,34	0,23	…	0,18
c_2	0,12	0,41	…	0,21
.	…	…	…	…
.	…	…	…	…
C_n	0,19	0,08	…	0,6

+

Likelihood matrix for Threat *(threat)*				
Nodes / Controls	c_1	c_2	…	c_n
v_1	-3	-6	…	-4
v_2	-1	-3	…	-3
.	…	…	…	…
.	…	…	…	…
v_n	-5	-2	…	-2

=

Resilience against *(threat)*				
Nodes / Controls	v_1	v_2	…	v_n
c_1	-2,15			
c_2				
.				
.				
C_n				

Figure 1.
Resilience security control calculation for the entire network against a single threat $\in \mathbb{T}$.

the time period over which a dependency between two infrastructures exhibits its maximum expected impact $I_{i,j}$, and $G_{i,j}$ denotes the expected growth of the failure. The growth rates used in this model are split into three types, namely: slow, linear or fast. Finally, let t denote an examined time period after a failure.

Growth rates $G_{i,j}$ are defined based on the maximum potential Impact $I_{i,j}$ and a growth relation between time step t and $T_{i,j}$. Specifically, "slow" growth rates follow a exponential evolution of type

$$I(t) = I^{\frac{t}{T}} \tag{5}$$

which begins at a slow pace and gradually increases in speed. "Linear" growth rates follow a typical approach

$$I(t) = I \cdot \frac{t}{T} \tag{6}$$

whereas "fast" impact growth rates are calculated using a logarithmic approach

$$I(t) = I \cdot \log_T t \tag{7}$$

in which incidents impose a very fast impact growth rate that gradually decreases in speed. For any $t > = T$, impact growth caps at $I(t) = I$.

In real-world implementations of the methodology, all aforementioned values for $T_{i,j}$ and $G_{i,j}$, along with $I_{i,j}$ and $L_{i,j}$, are obtained through on-site assessment, expert knowledge and quantification of infrastructure characteristics.

3.5 Qualitative ranking scales

The above equations need some sort of value ranges in order to quantify results. To support calculation of these equations, we opted to use the same scales as in [15]. All the values are assigned from the following Likert scales:

- $I\epsilon[1..9]$, where 1 is the lowest impact and 9 is the highest impact.
- $T, t\epsilon[1..10]$, which is a granular time scale that uses the unavailability time periods: 1 = 15 min, 2 = 1 h, 3 = 3 h, 4 = 12 h, 5 = 24 h, 6 = 48 h, 7 = 1w, 8 = 2w, 9 = 4w and 10 = more than 4w.
- $G\epsilon[1..3]$, where the value of 1 represents the slow growth rate, and values (2) and (3) represent the linear and fast evolution rates for impact respectively.

Each Impact value reflects a different qualitative criterion, based on the needs and threats of any given infrastructure. Nevertheless, quantification is uniform amongst all possible implementations, where a value of 1 reflects minimum to no Impact, while a value of 9 reflects catastrophic impact of an incident.

4. Case study: City of Rome

The city center of Rome was chosen as a case study due to the high concentration of various commercial activities and power centres both local and international as well as the presence of CIs which are essential to maintain vital societal functions (**Figures 2** and **3**). In particular, the area of interest holds the major Italian government offices, *San Giovanni Calibita Fatebenefratelli Hospital* located in the Tiber Island and *Termini Railway Station*, one of the most important railway stations of Italy as it connects Northern and Southern Italy.

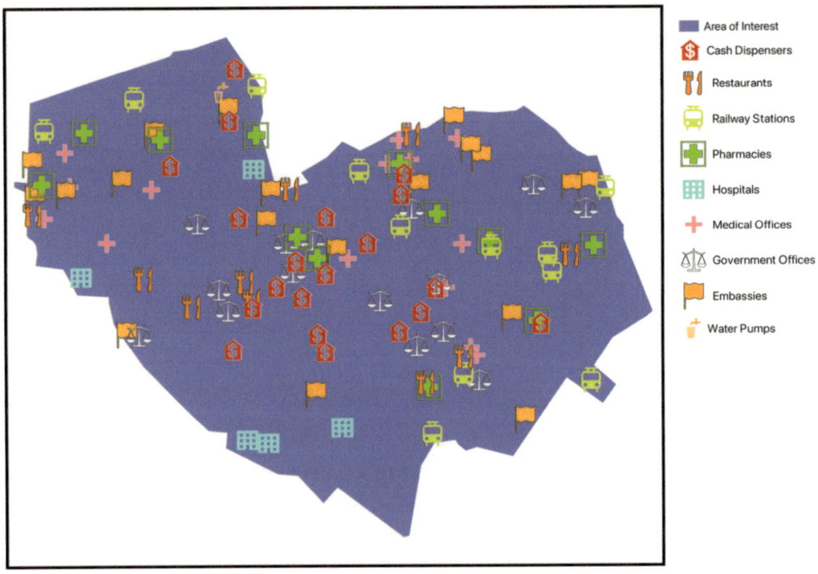

Figure 2.
The area of interest: an urban district of Rome. The map was anonymized and MV Electric substations and Base Transceiver Stations were removed to hide sensitive information.

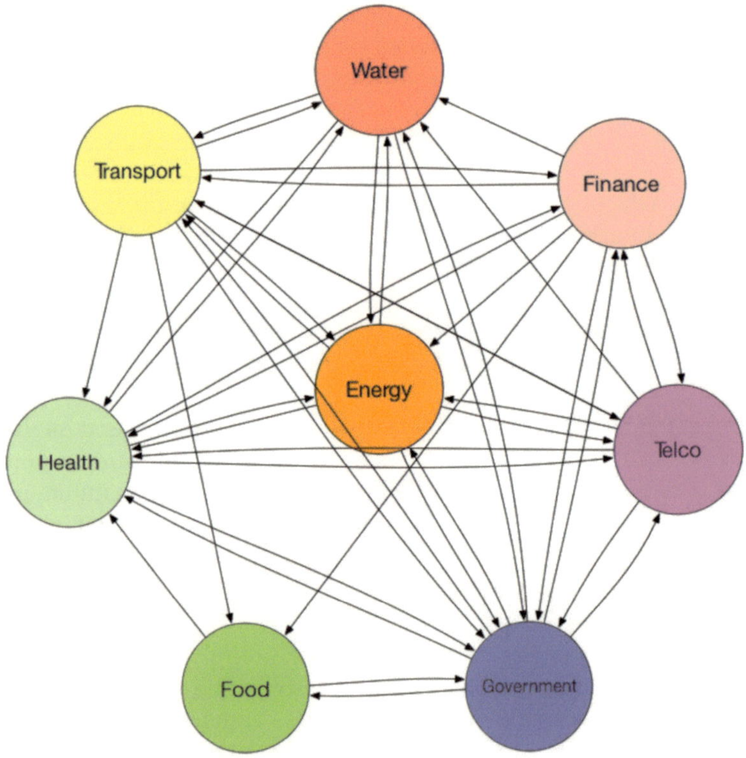

Figure 3.
The dependency graph used in the case study.

As reported in **Table 1**, we considered 8 categories including CI and Point of Interests and selected a set of specific components (nodes, hereafter) for each category that are located in the area of interest. In particular, we considered the following categories:

i. the Electric Distribution Network (EDN) of Rome consisting of 40 Medium Voltage (15 kV) substations;

ii. the Mobile Telecommunication System consisting of 31 Base Transceiver Stations (BTS);

iii. the Water Supply Network (WSN) consisting of 1 water pumping station;

iv. the Railway system including 12 stations;

v. a set of hospitals, medical offices and pharmacies;

vi. a set of government offices and embassies;

vii. a set of cash dispensers;

viii. a set of restaurants.

4.1 Dependency graph

In order to model the interdependencies among the different nodes, we assumed a cyber risk assessment as the case scenario. In particular, we considered a *dependency matrix* [26] that allows to reveal the potential vulnerability of a given node to the unavailability, corruption or disclosure of data from an interdependent node regardless of the current state of the shared data infrastructure. In other words, we assume a cyber threat $threat \in \mathbb{T}$ affecting the considered nodes and we use a *precomputed* dependency matrix as a means to assign a cyber vulnerability to each node w.r.t. the data disruption from all interdependent nodes.

Category	Subcategory	Acronym	Nr.
Energy	MV Electric substation	ES	40
Telecommunications	Base Transceiver Station	BTS	31
Finance	Cash Dispenser	CD	20
Government	Government Office	GO	15
	Embassy	EM	20
Transport	Railway Station	RS	12
Health	Medical Office	DO	15
	Pharmacy	PH	12
	Hospital	HP	5
Food	Restaurant	RE	10
Water	Water Pumping station	WP	1
Total:			182

Table 1.
CI categories and components modeled in the case study.

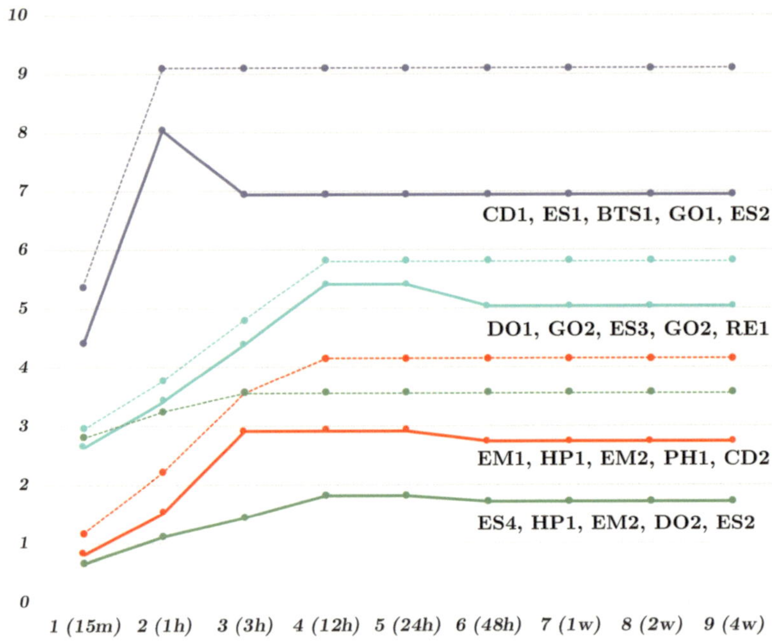

Figure 4.
A set of dependency risk paths with cumulative dependency risk. Dashed/continuous lines indicate the risk without/with the implementation of security controls.

The dependency matrix is consistent with the main cyber interdependencies that exist among the nodes modelled in the scenario although only a limited number of CI were considered for each sector present in the dependency matrix. Indeed, the electric substations (ES) supply energy to all nodes of other CI and thus a failure occurring in an ES would be disruptive for all nodes that receive energy from that ES. In addition, some of the ES are Remotely controlled and thus a failure occurring in those BTS nodes that in turn provide telecommunication services to the Remotely Controlled ES may compromise the control operations of the EDN.

In the absence of information regarding specific interdependencies, we employed a proximity criterion to model the relations among specific nodes. For example, we assumed that each energy consumer (i.e., all nodes that are not ES) is supplied by the nearest ES as well as each internet/telephony consumer is supplied by the nearest BTS. In addition, we did not model the intra-sector dependencies i.e. any dependency among the nodes of the same CI sector was not considered.

4.2 Likelihood matrix

As described previously, we employed the dependency matrix defined in [26] to model the interdependencies of the case study. That matrix was filled by gathering over 4.000 distinct data dependency metrics from CI stakeholders and reports the same CI sectors that were modelled in the case study and the cyber vulnerability of each sector w.r.t. all CI sectors. **Table 2** shows the value for both Inbound and Outbound data dependencies. Inbound data dependency represents information and data consumed by the examined CIs, while outbound data dependency represents the data leaving each examined CI, to be used by other CIs.

The columns for each sector represent how that sector is dependent by data coming into that sector. Most organisations can intuitively estimate this value, and that's how the data was collected in [26]. For example, in **Table 2**, column *BTS* represents the data, informations and services any BTS station would receive from each other sector, and how much that BTS station depends from that data, information or service.

Based on this matrix, we normalised the values and neglected the intradependencies and the low intradependencies. In other words, we treated the cyber vulnerability of a node as a likelihood that the node being affected. The resulting matrix is shown in **Table 2.**

4.3 Security Controls

Given the absence of information regarding the security controls implemented by the considered nodes, we assumed that each node v_i having a dependency with v_j where $j \in \{1, .., N_i\}$, is equipped with l^{v_i} security controls against the examined *threat*. We assumed that the likelihood values of the restoration controls $\overline{L}_{v_i,v_j} = const. \forall j \in \{1, .., N_i\}$. **Table 3** shows the likelihood values of the restoration controls.

4.4 Impact Assessment Criteria

In order to assess the impact of cyber attacks on the nodes, we considered the work of Fekete [27] that defines three impact assessment criteria in terms of critical proportion, time and quality aspects. Critical proportion refers to the number of elements or nodes of a CI such as critical number of services, size of population or number of customers affected and redundancies. Critical time considers aspects such as duration of outage, Mean Time to Repair (MTTR), Mean Time to Functionality (MTTF) and business continuity or interruption. Critical quality refers to the quality of the services delivered (e.g., the water quality) or the public trust in quality (e.g., trust in finance, feeling of security).

In the following subsections, a description of how the mentioned impact assessment criteria were applied to the case study will be provided. In particular, the assumptions that were made to take into account such criteria will be described in order to model the expected time-related impact $I(t)$ in terms of the maximum expected impact I, the impact time T and the impact growth rate G, as defined in Section 3.

	Inbound Dependencies										
CI Sector	**ES**	**BTS**	**CD**	**GO**	**EM**	**RS**	**DO**	**PH**	**HP**	**RE**	**WP**
ES	—	0.36	—	0.34	0.34	0.43	0.39	0.39	0.39	—	0.31
BTS	0.7	—	0.45	0.4	0.4	0.44	0.51	0.51	0.51	—	0.34
CD	0.71	0.72	—	0.4	0.4	0.4	0.42	0.42	0.42	0.44	0.5
GO	0.59	0.51	0.7	—	—	0.36	0.61	0.61	0.61	0.36	0.51
EM	0.59	0.51	0.7	—	—	0.36	0.61	0.61	0.61	0.36	0.51
RS	0.68	0.4	0.42	0.29	0.29	—	0.5	0.5	0.5	0.51	0.3
DO	0.41	—	0.3	0.51	0.51	—	—	—	—	—	0.44
PH	0.41	—	0.3	0.51	0.51	—	—	—	—	—	0.44
HP	0.41	—	0.3	0.51	0.51	—	—	—	—	—	0.44
RE	—	—	—	0.27	0.27	—	0.38	0.38	0.38	—	—
WP	0.49	—	—	0.29	0.29	0.32	0.36	0.36	0.36	—	—

Table 2.
The likelihood matrix used in the case study.

v_i	$\overline{L}_{v_i,v_j}$
ES, BTS, CD, GO, EM	0.3
RS, HP, WP	0.1
DO, PH, RE	0

Table 3.
Resilience influence of security control c^{v_i} on node v_j for the given threat with dependency risk subchain $v_i \to v_j$.

4.4.1 Maximum expected impact matrix

In order to apply the critical proportion criterion, given the difficulty of obtaining the number of customers supplied by a specific node from the CI owners, we assumed the number of inhabitants living in the geographical area where the specific node is located as the number of customers. Indeed, the areas considered are the census areas delivered by the *Italian National Institute of Statistics* (ISTAT) of which the number of inhabitants is known. This criterion was applied to model the maximum expected impact I for each couple of nodes i and j belonging to Energy, Telecommunication, Transport and Finance sectors. Thus, I was computed by combining the total number of customers supplied by i and j nodes so that the more customers are involved in the disruption of the nodes, the more impact we obtain.

Furthermore, the critical quality criterion was applied to compute I for each couple of nodes i and j belonging to Government, Health, Food and Water. In this case, we set a subjective value that takes into account the importance of the unavailability of the data for the specific nodes.

Table 4 summarises the criteria applied based on the sector nodes considered. It should be noticed that while I is time dependent when considering ES, BTS, RS and CD nodes (case A), this is not true when considering GO, EM, DO,PH, HP and RE nodes (case B) where I was set higher for the nodes that could be more impacted by the lack of data services. For case C, the two criteria were both considered and I was computed according to the metric reported in **Table 4**. The resulting impact matrix is shown on **Table 5**.

Let $v_0, v_1, .., v_n$ be a subchain of risk. We assumed that the reduction of impact $\overline{I}_{v_{i-1},v_i}$ on node v_i due to the restoration action $c^{v_{i-1}}$ implemented by v_{i-1} is given by:

$$\overline{I}_{v_{i-1},v_i} = \alpha \cdot I_{v_{i-1},v_i} \tag{8}$$

Table 6 shows the percentage of reduction α of the initial estimated damage I_{v_{i-1},v_i} for the generic dependency risk subchain $v_{i-1} \to v_i$.

4.4.2 Impact time and Impact growth rate matrices

Regarding the critical time criterion, we considered the expected duration of failure of nodes to compute the impact and growth time matrices. In particular, we assigned a low value to sectors that are highly dependent on the data availability and

Case	v_j	Impact assessment criterion	I_{v_i,v_j}
A	ES, BTS, RS, CD	Nr. of customers	node-dependent
B	GO, EM, DO, PH, HP, RE	Service criticality	sector-dependent

Table 4.
Maximum expected impact criteria for the dependency risk subchain $v_{i-1} \to v_i$.

	Inbound dependencies										
CI Sector	**ES**	**BTS**	**CD**	**GO**	**EM**	**RS**	**DO**	**PH**	**HP**	**RE**	**WP**
ES	—	⋆	—	7	4	⋆	4	4	7	—	7
BTS	⋆	—	⋆	7	4	⋆	3	3	6	—	6
CD	⋆	⋆	—	3	2	2	2	2	4	2	2
GO	8	8	3	—	—	⋆	3	3	5	3	5
EM	4	4	2	—	—	3	2	2	4	2	4
RS	⋆	⋆	2	⋆	3	—	3	3	4	3	3
DO	2	—	2	3	2	—	—	—	—	—	3
PH	2	—	2	3	2	—	—	—	—	—	3
HP	7	—	4	5	4	—	—	—	—	—	5
RE	—	—	—	3	2	—	2	2	2	—	—
WP	3	—	—	3	3	3	3	3	3	—	—

Table 5.
Maximum expected impact matrix used in the case study. ⋆ represents node-dependent impact.

v_{i-1}	v_i	α
ES, BTS	any	0.5
CD, GO, EM, RS, DO, PH, HP, RE, WP	any	1

Table 6.
Percentage of reduction α of the initial estimated damage I_{v_{i-1},v_i} for the dependency risk subchain $v_{i-1} \rightarrow v_i$.

that produce a quick impact such as Energy and Telecommunication and Finance and assigning a higher value to other sectors such as Water and Food that produce their negative effect in a longer period. The resulting impact time matrix is shown on **Table 7**.

Regarding the recovery time matrix, we modeled a time $\overline{T} = 15m$ for the electric substations ES are remotely controled as the SCADA system of the electric network allows to reactivate the electric supply in the order of minutes whereas $\overline{T} = 1h$ for a generic ES only a manual intervention performed by a repair crew can be operated with a longer time (approximately 1 hour). The resulting recovery time matrix is shown on **Table 8**.

Regarding the impact growth rate, **Table 9** shows the the criterion adopted and **Table 10** shows the resulting values for each couple of nodes. We considered the same growth rate for the recovery actions.

4.5 Results

The execution of the model based on the graph of 182 nodes produced about 750.000 risk paths with order ranging from five to eight and potential risk values between 0.27 and 9.53. **Figure 4** shows some significant dependency paths together with their cumulative dependency risk values.

The charts show that one dependency path (CD_1-ES_1-BTS_1-GO_1-ES_2) exhibits its highest risk value at time $t = 1h$ and then the implementation of mitigation strategies with a rapid response decreases the overall dependency risk. In general,

	Inbound dependencies										
CI Sector	**ES**	**BTS**	**CD**	**GO**	**EM**	**RS**	**DO**	**PH**	**HP**	**RE**	**WP**
ES	—	3 h	—	3 h	3 h	3 h	3 h	3 h	3 h	—	24 h
BTS	3 h	—	1 h	3 h	3 h	3 h	3 h	3 h	3 h	—	3 h
CD	3 h	3 h	—	3 h	3 h	3 h	12 h	12 h	3 h	2w	24 h
GO	3 h	3 h	3 h	—	—	12 h	12 h	12 h	12 h	2w	24 h
EM	3 h	3 h	3 h	—	—	12 h	12 h	12 h	12 h	2w	24 h
RS	3 h	3 h	3 h	12 h	12 h	—	12 h	12 h	12 h	2w	24 h
DO	3 h	—	3 h	24 h	24 h	—	—	—	—	—	24 h
PH	3 h	—	3 h	24 h	24 h	—	—	—	—	—	24 h
HP	3 h	—	3 h	24 h	24 h	—	—	—	—	—	24 h
RE	—	—	—	2w	2w	—	2w	2w	2w	—	—
WP	24 h	—	—	24 h	24 h	24 h	24 h	24 h	24 h	—	—

Table 7.
The maximum impact time matrix used in the case study.

	Inbound dependencies										
CI Sector	**ES**	**BTS**	**CD**	**GO**	**EM**	**RS**	**DO**	**PH**	**HP**	**RE**	**WP**
ES	—	15 m	—	15 m	15 m	15 m	15 m	15 m	15 m	—	15 m
BTS	3 h	—	1 h	1 h	1 h	1 h	1 h	1 h	1 h	—	1 h
CD	3 h	3 h	—	3 h	3 h	3 h	12 h	12 h	3 h	2w	24 h
GO	3 h	3 h	3 h	—	—	12 h	12 h	12 h	12 h	2w	24 h
EM	3 h	3 h	3 h	—	—	12 h	12 h	12 h	12 h	2w	24 h
RS	3 h	3 h	3 h	12 h	12 h	—	12 h	12 h	12 h	2w	24 h
DO	3 h	—	3 h	24 h	24 h	—	—	—	—	—	24 h
PH	3 h	—	3 h	24 h	24 h	—	—	—	—	—	24 h
HP	3 h	—	3 h	24 h	24 h	—	—	—	—	—	24 h
RE	—	—	—	2w	2w	—	2w	2w	2w	—	—
WP	24 h	—	—	24 h	24 h	24 h	24 h	24 h	24 h	—	—

Table 8.
The maximum recovery time matrix used in the case study.

		Growth rate node *i*		
G		**Slow**	**Linear**	**Fast**
Growth rate node *j*	**Slow**	Slow	Slow	Linear
	Linear	Slow	Linear	Fast
	Fast	Linear	Fast	Fast

Table 9.
Impact growth rate metric.

	Inbound dependencies										
CI Sector	**ES**	**BTS**	**CD**	**GO**	**EM**	**RS**	**DO**	**PH**	**HP**	**RE**	**WP**
ES	—	F	—	F	F	F	F	F	F	—	L
BTS	F	—	L	L	L	L	L	L	L	—	S
CD	F	L	—	L	L	L	L	L	L	L	S
GO	F	L	L	—	—	L	L	L	L	L	S
EM	F	L	L	—	—	L	L	L	L	L	S
RS	F	L	L	L	L	—	L	L	L	L	S
DO	F	—	L	L	L	—	—	—	—	—	S
PH	F	—	L	L	L	—	—	—	—	—	S
HP	F	—	L	L	L	—	—	—	—	—	S
RE	—	—	—	S	S	—	S	S	S	—	—
WP	L	—	—	S	S	S	S	S	S	—	—

Table 10.
The impact growth rate matrix used in the case study.

we observed an high risk value of subchains including the electric nodes due both to the high number of dependencies of nodes on the electric nodes and the high maximum impact associated.

Figure 5 shows a map representation of the dependency risk paths considered in **Figure 4** with the census areas involved. In particular, let $CA_1, CA_2, .., CA_M$ be the set of generic census area containing the CI nodes of all possible dependency chains. The generic CA_k s.t. $1 \leq k \leq M$, $CA_k = \{v_j\}$, $|CA_k| \leq n$ is associated specific a color according to the cumulative risk value $DR^k_{v_0,\dots,v_n}$ of a $v_0, v_1, .., v_n$ dependency subchain s.t. $\nexists$ a $p_0, p_1, .., p_g$ dependency chain s.t. $DR^k_{v_0,\dots,v_n} < DR^k_{p_0,\dots,p_g}$ with some

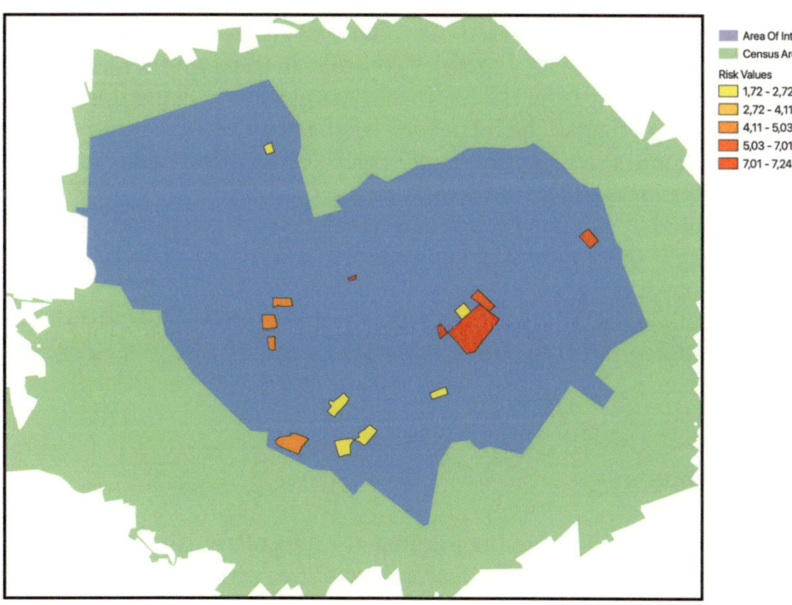

Figure 5.
Result map showing the risk value of each census area.

$p_h \in CA_k$ $(0 \leq h \leq g)$. In other words, each census area is colored according to the maximum risk value of a subchain that includes some nodes v_j that are located in that area (i.e. $v_j \in CA_k$).

Results depicted in **Figure 4** indicate cascading events between infrastructures. Each one of the four scenarios was validated to be true against real world data and historical analysis of such infrastructures. Following this, results indicate that the presented methodology is able to both (i) effectively project adverse effects from cascading events and accurately predict potential impact over time periods, and also (ii) highlight direct and indirect dependency vulnerabilities between highly dependent CIs.

On the latter, results delineate the criticality behind dependencies of Telecommunications and the Electrical sector. The sharp increase in impact over a very short time period (purple line, scenario 1) clearly shows that potential unavailability of the Electrical sector quickly and critically affects the Telecommunications. We followed up on this finding and results are proven true both from empirical analysis and also from historical data on locations analyzed by the tool.

Another potential use of the presented methodology includes capturing the effect of applying security controls and how these controls affect the resilience of systems over time. By analyzing the impact escalation and trajectory in analyzed attack paths, we see that the level of risk reduction for each of the presented scenarios is directly related with the time of deployment. Early application of security controls (scenario CD1, ES1, BTS1, GO1, ES2) seems to reduce the overall risk by 25% in less than two hours after the initiation of the attack path, while controls implemented later during the exposure to the adverse event show relatively smaller mitigation percentages of the overall risk (around 18%).

Red areas shown in **Figure 5** are highly populated areas containing electric nodes thus producing possible high impact in case of failure. This explains why several nodes of the subchains with high cumulative dependency risk are concentrated in this area.

5. Conclusions

By extending previous time-based dependency analysis models and by integrating the effect of resilience-related security controls, in this paper we have examined the effect of possible mitigation strategies in dynamically reducing the consequences of cascading effects. The model was applied to a real case study involving an urban area of Rome where a number of critical infrastructures deliver services to inhabitants and businesses. The model was set up by considering a precomputed dependency graph that exhibits the cyber dependencies of a set of infrastructures. The results highlight the most critical dependency chains and the areas with high concentration of critical nodes. The model was integrated into CIPCast Decision Support System allowing all actors involved in securing critical infrastructures to plan mitigation strategies aiming at reducing the overall risk of service degradation in the considered area.

Acknowledgements

Authors wish to acknowledge the funding of project RAFAEL (MIUR ARS01_00305) which has partly funded the research activities carried out for this work.

Author details

Vittorio Rosato[1]*, Antonio Di Pietro[1], Panayiotis Kotzanikolaou[2], George Stergiopoulos[3] and Giulio Smedile[4]

1 Laboratory for Analysis and Protection of Critical Infrastructures, Enea, Casaccia Research Centre, Rome, Italy

2 Department of Informatics, University of Piraeus, Greece

3 Department of Information and Communication Systems Engineering, University of Aegean, Samos, Greece

4 Degree in Informatics Engineering, Rome Tre University, Rome, Italy

*Address all correspondence to: vittorio.rosato@enea.it

IntechOpen

© 2021 The Author(s). Licensee IntechOpen. This chapter is distributed under the terms of the Creative Commons Attribution License (http://creativecommons.org/licenses/by/3.0), which permits unrestricted use, distribution, and reproduction in any medium, provided the original work is properly cited.

References

[1] A. Tofani, G. D'Agostino, A. Di Pietro, S. Giovinazzi, M. Pollino, and V. Rosato. Operational resilience: Concepts, design and analysis. *Special Issue "Emerging Approaches to Secure and Protect Critical Infrastructures", MDPI, Submitted*, 09 2020.

[2] A. Tofani, G. D'Agostino, A. Di Pietro, S. Giovinazzi, L. La Porta, G. Parmendola, M. Pollino, and V. Rosato. Modeling resilience in electrical distribution networks. In Samad M.E. Sepasgozar, Faham Tahmasebinia, and Sara Shirowzhan, editors, *Infrastructure Management and Construction*, chapter 3. IntechOpen, Rijeka, 2020.

[3] Alberto Tofani, Gregorio D'Agostino, Antonio Di Pietro, Giacomo Onori, Maurizio Pollino, Silvio Alessandroni, and Vittorio Rosato. Operational resilience metrics for a complex electrical network. In Gregorio D'Agostino and Antonio Scala, editors, *Critical Information Infrastructures Security*, pages 60–71, Cham, 2018. Springer International Publishing.

[4] Steven M Rinaldi, James P Peerenboom, and Terrence K Kelly. Identifying, understanding, and analyzing critical infrastructure interdependencies. IEEE control systems magazine, 21(6):11–25, 2001.

[5] Steven M Rinaldi. Modeling and simulating critical infrastructures and their interdependencies. In *37th Annual Hawaii International Conference on System Sciences, 2004. Proceedings of the*, pages 8–pp. IEEE, 2004.

[6] Michel Van Eeten, Albert Nieuwenhuijs, Eric Luiijf, Marieke Klaver, and Edite Cruz. The state and the threat of cascading failure across critical infrastructures: the implications of empirical evidence from media incident reports. Public Administration, 89(2):381–400, 2011.

[7] Enrico Zio and Giovanni Sansavini. Modeling interdependent network systems for identifying cascade-safe operating margins. IEEE Transactions on Reliability, 60(1):94–101, 2011.

[8] J Talsma, B Becker, Quanduo Gao, and ERIK Ruijgh. Coupling of multiple channel flow models with openmi. In *Proceedings of the Tenth International Conference on Hydroinformatics*, 2012.

[9] Serge P Hoogendoorn and Piet HL Bovy. State-of-the-art of vehicular traffic flow modelling. *Proceedings of the Institution of Mechanical Engineers, Part I: Journal of Systems and Control Engineering*, 215(4):283–303, 2001.

[10] Samitha Samaranayake, Sébastien Blandin, and Alexandre Bayen. Learning the dependency structure of highway networks for traffic forecast. In *2011 50th IEEE Conference on Decision and Control and European Control Conference*, pages 5983–5988. IEEE, 2011.

[11] Mohammad Shahraeini and Panayiotis Kotzanikolaou. A dependency analysis model for resilient wide area measurement systems in smart grid. IEEE Journal on Selected Areas in Communications, 38(1):156–168, 2019.

[12] Min Ouyang and Leonardo Dueñas-Osorio. An approach to design interface topologies across interdependent urban infrastructure systems. Reliability Engineering & System Safety, 96(11): 1462–1473, 2011.

[13] Erich Rome, Sandro Bologna, Erol Gelenbe, Eric Luiijf, and Vincenzo Masucci. Diesis: an interoperable european federated simulation network for critical infrastructures. In *Proceedings of the 2009 SISO European Simulation Interoperability Workshop*, pages 139–146, 2009.

[14] Christos Siaterlis, Bela Genge, and Marc Hohenadel. Epic: a testbed for scientifically rigorous cyber-physical security experimentation. IEEE Transactions on Emerging Topics in Computing, 1(2):319–330, 2013.

[15] George Stergiopoulos, Panayiotis Kotzanikolaou, Marianthi Theocharidou, Georgia Lykou, and Dimitris Gritzalis. Time-based critical infrastructure dependency analysis for large-scale and cross-sectoral failures. International Journal of Critical Infrastructure Protection, 12:46–60, 2016.

[16] George Stergiopoulos, Panayiotis Kotzanikolaou, Marianthi Theocharidou, and Dimitris Gritzalis. Risk mitigation strategies for critical infrastructures based on graph centrality analysis. International Journal of Critical Infrastructure Protection, 10: 34–44, 2015.

[17] Adam Hahn and Manimaran Govindarasu. Smart grid cybersecurity exposure analysis and evalution framework. In *IEEE PES General Meeting*, pages 1–6. IEEE, 2010.

[18] Sumeet Jauhar, Binbin Chen, William G Temple, Xinshu Dong, Zbigniew Kalbarczyk, William H Sanders, and David M Nicol. Model-based cybersecurity assessment with nescor smart grid failure scenarios. In *2015 IEEE 21st Pacific Rim International Symposium on Dependable Computing (PRDC)*, pages 319–324. IEEE, 2015.

[19] Earl E Lee II, John E Mitchell, and William A Wallace. Restoration of services in interdependent infrastructure systems: A network flows approach. *IEEE Transactions on Systems, Man, and Cybernetics, Part C (Applications and Reviews)*, 37(6):1303–1317, 2007.

[20] Nils K Svendsen and Stephen D Wolthusen. Analysis and statistical properties of critical infrastructure interdependency multiflow models. In *2007 IEEE SMC Information Assurance and Security Workshop*, pages 247–254. IEEE, 2007.

[21] Vittorio Rosato, Limor Issacharoff, Fabio Tiriticco, Sandro Meloni, S Porcellinis, and Roberto Setola. Modelling interdependent infrastructures using interacting dynamical models. *International Journal of Critical Infrastructures*, 4(1-2):63–79, 2008.

[22] White House. *Critical infrastructure security and resilience*. White House, 2013.

[23] Panayiotis Kotzanikolaou, Marianthi Theoharidou, and Dimitris Gritzalis. Assessing n-order dependencies between critical infrastructures. *International Journal of Critical Infrastructures* 6, 9(1-2):93–110, 2013.

[24] Panayiotis Kotzanikolaou, Marianthi Theoharidou, and Dimitris Gritzalis. Cascading effects of common-cause failures in critical infrastructures. In *International Conference on Critical Infrastructure Protection*, pages 171–182. Springer, 2013.

[25] Panayiotis Kotzanikolaou, Marianthi Theoharidou, and Dimitris Gritzalis. Interdependencies between critical infrastructures: Analyzing the risk of cascading effects. In *International Workshop on Critical Information Infrastructures Security*, pages 104–115. Springer, 2011.

[26] T. Macaulay. *Critical Infrastructure: Understanding Its Component Parts, Vulnerabilities, Operating Risks, and Interdependencies*. Taylor & Francis, 2008.

[27] Alexander Fekete. Common criteria for the assessment of critical infrastructures. *International Journal of Disaster Risk Science*, 2:15–24, 03 2011.

Chapter 6

A Strategy to Improve Infrastructure Survivability via Prioritizing Critical Nodes Protection

Luca Faramondi, Giacomo Assenza, Gabriele Oliva, Ernesto Del Prete, Fabio Pera and Roberto Setola

Abstract

From an engineering point of view, the survivability of a system is defined as its ability to continue to operate despite a natural or human-made disturbance; for example a serious mechanical fault, a human error, or a malicious cyber or physical attack. In the context of critical infrastructures, due to their relevance for the public wellness, it is mandatory to improve the robustness of such systems in order to ensure the availability of essential services such as the distribution of water, gas and electrical power. Nowadays, due to the increasing number of cyber incidents, the definition of protection strategies, able to improve the survivability level of this infrastructure, is at the heart of the scientific debate. In this chapter we propose a procedure based on three steps aimed at improving infrastructure survivability. In the first stage we propose some approaches to identify the criticality degree of each subsystem composing the infrastructure, in the second stage we propose a method to aggregate multiple criticality evaluations performed by subject matter experts by providing a unique holistic indicator. Finally, on the basis of such indicator, we propose a protection strategy to improve the robustness of the entire system.

Keywords: critical nodes, network robustness, protection strategy, optimization problem, cooperative games

1. Introduction

The physical and cyber protection of critical infrastructures (CIs) is crucial to ensure the availability of multiple essential services. Concerning the physical security aspects, critical infrastructures are, in most cases, complex and geographically distributed systems hence hard to protect. Regardless of the specific scenario, a CI can be represented as a set of sub-systems able to interact and cooperate in order to provide services that are essential for the economy, society and public wellness. For example, in gas distribution systems, the cooperation of metering and regulation stations is fundamental to guarantee the proper functioning of the entire infrastructure. In power grids and water distribution infrastructures, the availability of

 IntechOpen

electrical power and water, depends respectively on the joint action of singular sub-systems such as bus or water supply stations. Analogously, the correct operation of a plant depends on the right operativeness of several elements as illustrated by the 4STER European project.

Critical infrastructure are characterized by a high level of interconnection and interdependency where the operation of a subsystem is essential for the functioning of others. In such a context, the disruption of a subsystem can easily escalate creating waterfall effect impacting multiple services and geographic areas. Therefore, in order to guarantee the functioning of the entire infrastructure it is necessary to protect adequately each sub-system from fault or exogenous events potentially capable of compromising normal operativity levels. As reported in [1], on the 28th September 2003, in Italy and some areas of Switzerland, about 56 million people lost power due to a storm-tossed tree branch that hit Swiss power lines. About 30,000 people remained trapped in trains, several hundred passengers were stranded on underground transit systems, and there were significant knock-on effects across other critical infrastructures. Similarly, the 2005 Hurricane Katrina [2] caused widespread power outages throughout Louisiana, Mississippi, Alabama, Florida, Kentucky and Tennessee due to the cascading effects initiated by a local event. Another example is the 2011 Great East Japan earthquake [3] and the resulting tsunami: 1.5 million households did not have access to their water supply, 4.4 million households were left without electricity, and all the local railway services were halted, and communications were suspended.

Domino effects over the entire infrastructure due to local fault are not caused only by accidental faults or natural disasters, but could also be intentionally caused by malicious actors. For example, with the increasing reliance of CI on Information & Communication Technology (ICT) malicious actors can perform attacks via cyberspace triggering service disruptions significant economic losses and even kinetic effects. This has been particular concerning in relation to the energetic sector with a significant increase of cyber threats capable of causing outages and blackout in power systems.

The first example of how a cyber attack can affect the operativity of CI causing mechanical damage was provided by the Aurora project [4]. This was a test performed by the Idaho National in which the simulation of a cyberattack led to the destruction of a 27-ton generator. Another Significant example is represented by the Stuxnet worm. The worm was able to modify the rotation speed of particular motors installed inside the centrifuges used for the uranium enrichment in plant in Iran. Similarly, recent blackouts in Ukraine in 2015 and 2016 were respectively caused by Blackenergy3 and CrashOverride, two malware specially designed to cause blackouts via cyber intrusion [5].

In addition, we have to consider impacts on workers' safety. Power plants, water plants, gas plants can provoke accidents and enormous damages. Seveso plants can be used for the storage of hazardous materials: an attack aimed at these plants can also cause a domino effect. The capability to adjust machine parameters in order to improve performance or simply in order to change behavior can make other people with criminal intent adjust parameters so that workers and others can be put at risk of harm. Example of parameters can be speeds, forces, torques that can be put at dangerous levels. In addition, graphical interfaces used for human-machine interaction can be altered so that people could see a situation not corresponding to reality (not reported error codes or messages, different values of parameters or measures). In order to identify hazards associated to the use of a machine or a set of machines, procedures like HAZOP, HAZID, accident reviews must be taken into consideration. Anyway, security and safety must be considered as part of the normal working processes and not always this happens.

The main common aspect about these cited events is that a local event is able to compromise the functionality of the entire plants due to a domino effect. The identification of the most critical sub-systems is a crucial point for the definition of effective protection strategies able to improve the survivability of the systems. To this end, it is fundamental to identify adequate metrics and indicators to quantify the criticality rate associated to each sub-system, especially in highly heterogeneous contexts.

1.1 Related works

From the literature, one of the typical strategies to obtain such metrics is to simulate the effects of negative events, such as local faults, in order to provide insights on the most critical elements, for which protection needs to be raised. In particular, a well-established approach is to focus on intentional attacks, considering a rational attacker that aims at maximizing the damage while keeping low the effort required for his/her malicious action. Starting from the seminal works of Arulsevan et al. [6] it has become paramount that attacks that take into account the topology of the infrastructure, can select more effectively the target sites, increasing the damage dealt (e.g., in terms of disconnection of large portions of the infrastructure by causing services interruption). In [7–10] multiple approaches for the identification of critical nodes in infrastructure networks are presented. All these methods consists in optimization problems able to discover the nodes whose removal from the network compromise the connectivity of the entire system. All these approaches requires initial assumptions about the attacker budget and preferences despite this information are not available in general in a real context. Moreover, the results of these approaches are able to highlight the most critical node in a network but not provide a metric capable of quantifying the degree of criticality for each node of the infrastructure. In more details, the approach presented in [7] proposes a method, able to identify the most critical nodes, based on the result of an optimization problem characterized by the presence of assumptions about the strategy of an attacker in terms of available budget and dimension of disconnected components. Similar assumptions are considered also in the approach presented in [8, 9], the authors propose a method which aims at minimizing the attack cost against the infrastructure with constraints about the features of the network. Finally, assumptions about the attacker preferences are also required in the formulation presented in [10]. In general, centrality measures, such as the node degree or betweenness centrality are often adopted as criticality measures, while in [11] the authors propose a critical index for the elements of a CI by analyzing the solutions of a multi objective optimization problem without any assumption about the attacker behaviour. However, the adoption of a unique metric or indicator about the criticality rate of each node of the system is quite unrealistic due to the complex nature of the infrastructures. Two approaches able to consider multiple metrics with the aim to compute a final aggregated criticality holistic indicator are presented in [12, 13]. The proposed approaches take into account multiple indicators based on multiple data source (topology data, field-related data, expert evaluations, etc.) but not provide a final step necessary to define a defensive strategy and evaluate its effectiveness.

1.2 Contribution and outline of the chapter

In this chapter we want to propose a procedure able to define a defensive strategy for CIs based on multiple node criticality measures. In more details, the procedure is based on three steps, as depicted in **Figure 1**: In the first stage (Section 2) we provide some specific criticality measure for CIs based on the connectivity of the system. The identification of the criticality measures is a fundamental stage in

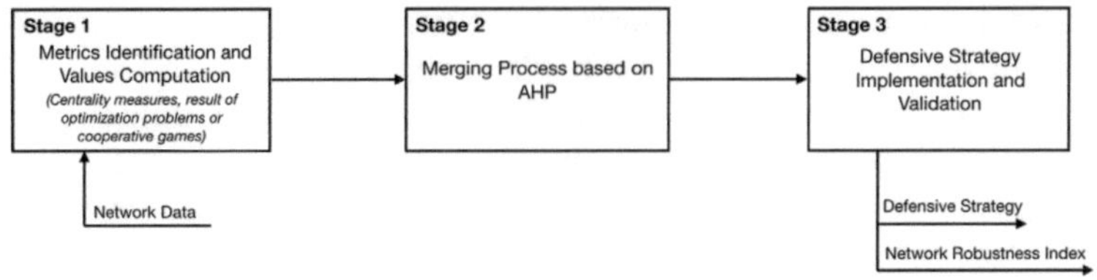

Figure 1.
Flow chart of the proposed three-steps procedure.

the defensive strategy definition process. In literature, graph centrality measures are often adopted as criticality measures for infrastructure but these approaches (e.g. Node degree or node betweenness) are quite ineffective as proved in [11]. In the second stage (Section 3) a methodology to merge multiple criticality metrics, based on the well-known Analytic Hierarchy Process [14], is described in order to overcome the limit about the application of a single metric in a complex environment. Moreover, such methodology allows considering also the criticality evaluations given for a subset of infrastructure nodes. The definition of the defensive strategy is provided in the last step (Section 4) and its effectiveness is proved by analyzing the global robustness of the network with respect to multiple robustness evaluation methods. Finally, in (Section 5), the application of the three-step procedure is illustrated with respect to the case study network with the aim of proving the effectiveness of the proposed strategy.

1.3 Notation

Let us denote by $|X|$ the cardinality of a set X; moreover, we represent vectors via boldface letters, and we use $\mathbf{k}_m$ to indicate a vector in $\mathbb{R}^m$ whose components are all equal to k, while by I_n we identify the $n \times n$ identity matrix. Finally, we denote the sign of $x \in \mathbb{R}$ by $sign(x)$ and by $sign(X)$ the entry-wise sign of a matrix X. Let $G = \{V, E\}$ denote a *graph* with a finite number n of nodes $v_i \in V$ and e edges $(v_i, v_j) \in E \subseteq V \times V$, from node v_i to node v_j. A graph is said to be *undirected* if $(v_i, v_j) \in E$ whenever $(v_j, v_i) \in E$ (see **Figure 2**). The *adjacency matrix* of a graph G is an $n \times n$ matrix A such that $A_{ij} = 1$ if $(v_j, v_i) \in E$ and $A_{ij} = 0$ otherwise. A *path* over an undirected graph $G = \{V, E\}$, starting at a node $v_i \in V$ and ending at a node

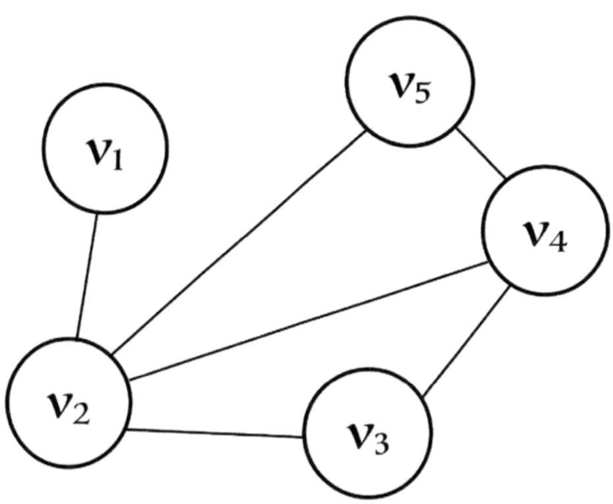

Figure 2.
Example of a graph $G = \{V, E\}$ with $|n| = 5$ nodes and $|E| = 6$ edges.

$v_j \in V$, is a subset of links in E that connects v_i and v_j without creating loops. An undirected graph $G = \{V, E\}$ is *connected* if each node can be reached by each other node by means of the links in E.

For the sake of clarity, we report here the notation adopted in the rest of the chapter.

c_i	Removal cost for node v_i
$PWC(G)$	Pairwise connectivity of G
$NPWC(A, x)$	Normalized pairwise connectivity for a graph with adjacency matrix A and without considering nodes v_i s.t.$\mathbf{x}_i = 0$
$\mathcal{P}$	Pareto Front
χ_i	Critical index for node v_i
P	Set of players in the cooperative game
$\Gamma(P, g)$	Cooperative game for players in P evaluated via characteristic function g
ϕ_i	Shapley value for player i
M_i	i-th metric
m	Number of metrics
$r_a^{(i)}/r_b^{(i)}$	Relative utility ratio among alternatives i and j according to metric i
R_{ab}^i	Matrix of utility ratios among alternatives a and b according to metric i
B	Defensive budget
w_i	Relevance of metric i
$\prod$	Global robustness index

2. Node criticality metrics based on network connectivity

As mentioned above the first step of the proposed approach for the identification of a defensive strategy is the identification of metrics of interests able to evaluate the network criticalities from multiple points of view. Despite in literature this process is often reduced to a simple centrality measure computation, in this section we propose two other applicable approaches, based on the infrastructure connectivity, to compute the criticality of each sub-systems of a CI. For the sake of clarity, in this context we represent the entire infrastructure via undirected graph $G = \{V, E\}$ where V is the set of n nodes v_i, (each node represents a sub-system of the CI) and $E \subseteq V \times V$ is the set of e undirected edges (v_i, v_j). An edge connects two nodes if a real physical connection exists between the two corresponding sub-systems.

Both the approaches for the critical node identification, presented in this section are based on the concept of connectivity. In our models, when a node is attacked and is unable to operate, we remove the node and the incident edges from the graph. The deletion of particular critical nodes could compromise the connectivity of the other elements of the network. Notice that, for each node v_i we consider a removal cost $c_i > 0$. With the aim to measure the degree of connectivity of the graph G, we adopt the Pairwise Connectivity (PWC), it is an index that captures the overall degree of connectivity of a graph on the basis of the couples of nodes connected by means of edges in G.

$$PWC(G) = \sum_{(v_i, v_j) \in V \times V, v_i \neq v_j} p(v_i, v_j), \tag{1}$$

where $p(v_i, v_j)$ is 1 if the pair (v_i, v_j) is connected via a path in G, and is zero otherwise. Noting that the maximum number of couples of nodes in a graph with n nodes is $\frac{n(n-1)}{2}$, the *normalized pairwise connectivity* (NPWC) is defined as:

$$NPWC(G) = \frac{2PWC(G)}{n(n-1)} \in [0, 1]. \tag{2}$$

Remark 1 $NPWC(G)$ is a measure of connectivity of the graph G, in fact, it is easy to note that

$$G \text{ connected} \Leftrightarrow NPWC(G) = 1. \tag{3}$$

When $NPWC(G) < 1$, the graph is not connected, but the larger $NPWC(G)$ is, the more G is "close" to be a connected graph. □

We now provide a more descriptive definition of a NPWC by taking into account a subset of attacked nodes. Let A be the adjacency matrix of an undirected graph $G = \{V, E\}$ and let $\mathbf{x} \in \mathbb{R}^n$ be a column vector whose entries $x_i = 0$ if the i-th node has been removed due to an attack or a fault and $x_i = 1$ otherwise, we define the connectivity as:

$$NPWC(A, \mathbf{x}) = \frac{\mathbf{1}_n^T \left[sign\left(\sum_{i=0}^{n-1} \hat{A}^i \right) - I_n \right] \mathbf{1}_n}{n(n-1)} \tag{4}$$

where $\hat{A}_{ij} = A_{ij} x_i x_j$, $\mathbf{1}_n$ is a column vector composed by n entries equal to 1.

2.1 A critical index based on optimization problem

The definition of the Critical Index χ_i for a node v_i, come directly from the solutions of a multi-objective problem defined by assuming the point of view of a malicious attacker.

In Eq. (5) the behavior of an attacker is defined as a multi-objective optimization problem characterized by two conflicting objectives: the reduction of the connectivity in terms of NPWC and the simultaneous minimization of the required attack effort in terms of removal cost. We reiterate that if an attacker want to disconnect a node v_i from the graph then (s)he must pay a cost c_i.

Problem 1

$$\min_{\mathbf{x} \in \{0,1\}^n} f(\mathbf{x}) = \min \left[f_1(\mathbf{x}), f_2(x) \right]^T, \tag{5}$$

where $\mathbf{x}$ represents the vector of decision variables, whose entries x_i are equal to 0 if the node v_i is involved in the attack, 1 otherwise and where

$$f_1 = NPWC(A, \mathbf{x}) \tag{6}$$

and

$$f_2 = \frac{\mathbf{c}^T (\mathbf{1}_n - \mathbf{x})}{\mathbf{1}^T \mathbf{c}} \tag{7}$$

where $\mathbf{c} = [c_1 \dots c_n]^T$ is the vector whose entries represent the cost necessary to remove each node from the graph.

As described in [11], in general, a multi-objective problem is characterized by the presence of multiple optimal solutions $\mathbf{x}^{(j)}$ collected in the Pareto front set $\mathcal{P}$. Each solution is associated to a couple of values $\left[f_1\left(\mathbf{x}^{(j)}\right), f_2\left(\mathbf{x}^{(j)}\right)\right]$ according to the two objective functions. In other words, each optimal solution $\mathbf{x}^{(j)}$ represents a different attack strategy with damages caused on the network $f_1\left(\mathbf{x}^{(j)}\right)$ and different attack effort $f_2\left(\mathbf{x}^{(j)}\right)$ as depicted in **Figure 3**.

In [11], the Critical Index χ_i is defined as in Eq. (8):

$$\chi_i = \frac{\sum_{\forall \mathbf{x}^{(j)} \in \mathcal{P}} \mathbf{x}_i^{(j)}}{|\mathcal{P}|} \tag{8}$$

where $|\mathcal{P}|$ represents the number of solutions in the Pareto front. In other words it is defined as the ratio between the frequency with which a node v_i is involved in the attacks listed in the Pareto front and its cardinality. If the critical index χ_i is close to 0 this implies that the node is rarely involved in attack plans, instead, the closer it is to 1, more frequently the node is involved in optimal attack strategies.

2.2 A critical index based on a cooperative game

An alternative approach for the identification of the most critical nodes in a network is presented in [15]. Analogously to the critical index based on the results of the multi-objective optimization problem, the proposed method is based on the concept of NPWC. Differently from the previous critical index, this measure come from the game theory and is based on the solution of a cooperative game.

A cooperative game, sometimes called a value game or a profit game, is a competition among groups of players. Formally, a cooperative game is defined by a set of players P and a characteristic function $v : 2^N \Rightarrow \mathbb{R}^+$ which associate to all possible coalitions of players a utility rate. The function describes how much collective payoff a set of players can gain by forming a coalition.

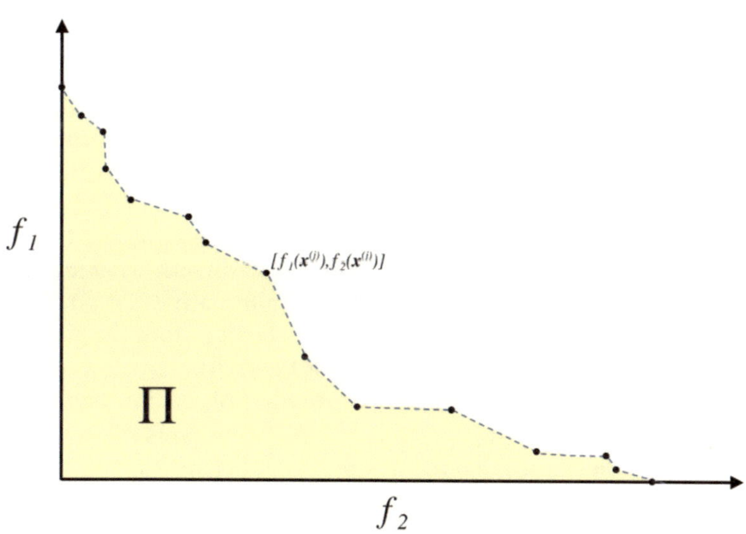

Figure 3.
Pareto front: the optimal solutions set for multi-objective optimization problem.

Let P be the set of players, and $g : 2^P \Rightarrow \mathbb{R}^+$ a function that satisfies the following properties:

- $g(\varnothing) = 0$
- Superadditive property: if $S, T \in 2^P$ *s.t.* $S \cap T = \varnothing$, then $v(S \cup T) \geq g(S) + g(T)$

The cooperative game $\Gamma(P,g)$ is defined by the couple (P,g) where the elements of P are the players of the game and the characteristic function of the game $g(S)$ estimates the utility of each coalition $S \in 2^P$.

Cooperative games can be solved via multiple approaches, the Shapley value [16] is one of the possible concepts of solution. The Shapley value assigns to each player $i \in P$, a reward ϕ_i. The larger is the contribution given by i in all the possible coalitions of players, based on the function g, the larger is the reward ϕ_i for the player i.

The Shapley value is a column vector Φ whose entries are ϕ_i are defined according to Eq. (9).

$$\phi_i = \frac{1}{n!} \sum_{S \subseteq P\{i\}} |S|!(n-|S|-1)!(g(S \cup \{i\}) - g(S)) \tag{9}$$

With the aim to adopt these concepts to provide a critical index able to quantify the criticality of each node of the network, a cooperative game $\Gamma(N, nPWC)$ is defined. The set of players is represented by the set of nodes N while the characteristic function g is *NPWC* (Eq. (3)). Notice that, in [15], it is demonstrated that the NPWC satisfy the two fundamental properties of a characteristic function.

The solution of the proposed game will assign a reward to each node in V proportional to its contribution to the connectivity expressed in terms of *NPWC*, hence the Shapley value can be considered a valid node criticality metric.

3. A multi-criteria vulnerability detection index

As briefly introduced in Section 1, a research of the most critical nodes based on a single metric is practically worthless and extremely simplistic. In this section we propose an approach able to provide a holistic indicator able to take into account multiple criticality evaluations based on multiple metrics also in presence of incomplete data. The proposed method is based on the well-known Analytic Hierarchy Process (AHP) introduced by Saaty [17]. For a given set of m alternatives, relative utility ratios r_i/r_j are defined by experts. Such a setting is typical in contexts involving human decision-makers, which are usually more comfortable providing relative comparisons among the utilities of the different alternatives (e.g., "Alternative i is twice better than alternative j"), rather than directly assessing an absolute utility value of each alternative (i..e, "The value of alternative i is ..."). The AHP is a procedure able to estimate the absolute utilities r_i starting from the given utility ratios r_i/r_j. See [17] for additional notions about the AHP.

We now suppose to have m different metrics $M_1 \ldots M_m$. According to these metrics, the entries of the column vectors $\mathbf{r}^{(1)} \ldots \mathbf{r}^{(m)}$ represents the criticality rate of each node of the graph. Notice that the method is applicable also if for some metrics the criticality ratio of some node is not available [12]. Finally, let $w_1 \ldots w_m$ be positive weights defined by subject-matter experts (SMEs) representing the relevance of each metric. The larger is the weight associated to the i-th metric, the larger the influence of such metric in the final holistic indicator. Such weights can be obtained also resolving AHP on the basis of pair-wise comparisons between the different metrics.

For each metric we define the $n \times n$ matrix $R^{(i)}$ whose entries are defined as follows:

$$R_{ab}^{(i)} = \begin{cases} r_a^{(i)}/r_b^{(i)} \text{if both} r_a^{(i)} \text{and} r_b^{(i)} \text{are defined} \\ 0 \; otherwise \end{cases} \tag{10}$$

In other words, the matrix $R^{(i)}$ collects the relative utility ratios between the a-th and b-th nodes according to the i-th metric if both the evaluation are available. Notice that some ratio $r_a^{(i)}/r_b^{(i)}$ might be undefined if $r_b^{(i)} = 0$, due to this reason, we treat zero-valued entries as not available data.

By considering the matrices $R^{(i)}$, we aim at finding the aggregated holistic indicator $\mathbf{r}^* \in \mathbb{R}^n$ that solves the following problem.

Problem 2 Find $\mathbf{r}^* \in \mathbb{R}^n$ *that solves*

$$\mathbf{r}^* = \arg\min_{\mathbf{r} \in \mathbb{R}_+^n} f(\mathbf{r}) = \sum_{i=1}^{m} w_i \sum_{a=1}^{n} \sum_{b \mid R_{ab}^{(i)} \neq 0} \left(\ln \left(R_{ab}^{(i)} \right) - log\,(r_a) + log\,(r_b) \right)^2 \tag{11}$$

The holistic indicator $\mathbf{r}^*$ is a new node criticality measure that represents a compromise between the m initial metrics $M_1 \dots M_m$ by taking into account the SMEs preferences w_i. In other words, Problem 2 aims at finding the criticality indicator $\mathbf{r}_a^*$, assigned to the a-th node, such that the ratios $\mathbf{r}_a^*/\mathbf{r}_b^*$ minimize the deviation from the ratios $R^{(i)}$ for the m considered metrics.

4. Defensive strategy definition and evaluation

In this section we propose a methodology to define a defensive strategy able to improve the survivability of the network with a focus on the connectivity maintenance with respect to nodes deletion. As introduced in Section 2, an attack cost c_i is associated to each node v_i. Our aim is the definition of a new distribution of the budget in order to minimize the loss of connectivity in case of malicious attacks.

Let $B = \sum_{i=1}^{n} c_i$ the defensive budget computed on the basis of the initial removal costs. We propose a new allocation of the budget by defining the removal cost proportionally to the holistic indicator $\mathbf{r}^*$ described in Section 3. Hence, we define the new removal costs $\hat{c}_i$ as follows:

$$\hat{c}_i = \frac{1}{B} \frac{\mathbf{r}^*{}_i}{\mathbf{1_n}^T \mathbf{r}^*}. \tag{12}$$

It is now necessary evaluate the robustness of a network with a particular defensive strategy. As introduced in Section 2.1, due to its multi-objective nature, Problem 1 is characterized by the presence of multiple optimal solutions collected in the Pareto front $\mathcal{P}$. Each optimal solution $\mathbf{x}^{(j)}$ is associated to a couple of values: a particular connectivity value $f_1(\mathbf{x}^{(j)})$ and an attack cost $f_2(\mathbf{x}^{(j)})$, where f_1 and f_2 represent the two objective function of Problem 1.

In [11], the global robustness index $\prod$ is defined as the area under the polygonal chain connecting the points $(f_1(\mathbf{x}^{(j)}), f_2(\mathbf{x}^{(j)}))$ in the Pareto front using trapezoidal rule for numerical integration.

As depicted in **Figure 3**, $\prod$ is a measure of the overall robustness of the network. In fact, the larger is the area, the higher is the value of the objectives associated to the solutions in the Pareto front; hence, high values of the global robustness index correspond to networks where the attacker is not able to deal large damage, or deals large damage only for large effort.

5. Case study

In this section we prove the effectiveness of the proposed three-stage methodology able to improve the network survivability via critical nodes protection. The proposed strategy is tested on the CI represented by the network depicted in **Figure 4**. Notice that the case study is based on a network that does not represents a real infrastructure. The network is composed by $n = 15$ nodes and $e = 35$ edges. As discussed in Section 2, the first step of the methodology is devoted to the identification of criticality measures able to take into account the effects about the disconnection of a node from the graph by evaluating the loss of connectivity of the entire infrastructure. Notice that, the removal costs c_i are set to 1 for each node of the infrastructure.

The first columns in **Table 1** collect the metrics defined by Eqs. 5 and 6 respectively. Concerning the distribution of the critical indices χ_i, the largest value are associated to the nodes 10 and 3. Notice that, the deletion of such nodes divides the nodes in two partitions, hence it strongly compromises the connectivity of the network in terms of nPWC.

Similar results are obtained by considering the computation of the Shapley value in order to solve the cooperative game as described in Section 2.2. We remark that this approach assigns a reward to each node of the network according to their contribution to the connectivity of the entire network by considering all the possible partitions of nodes. Notice that, the results computed via Shapley not consider the removal cost c_i while the results of Problem 1 take into account also this aspect, moreover, in this case study all the removal costs c_i are set to 1.

Finally, the fourth and fifth columns of **Table 1** collect the node degree and the betweenness centrality [18] for each node in the graph.

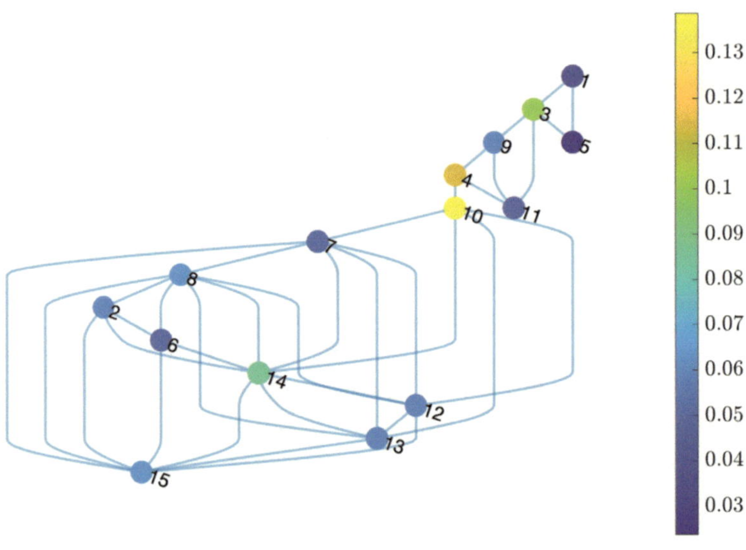

Figure 4.
Case study network. The node color depends on the holistic criticality rate computed via Eq. (7).

Node	Critical Index χ_i[a]	Shapley Value ϕ_i[b]	Degree	Betweenness	Holistic Indicator r_i^*[c]
1	0.1111	0.0354	2	0	0.0360
2	0.1852	0.0493	4	0	0.0587
3	0.2863	0.0952	4	24	0.1003
4	0.2593	0.1465	3	45	0.1143
5	0.0370	0.0354	2	0	0.0238
6	0.1111	0.0493	4	0	0.0484
7	0.1111	0.0521	6	3.5	0.0491
8	0.2593	0.0547	7	2	0.0618
9	0.1481	0.0515	3	15	0.0580
10	0.2963	0.1635	5	48	0.1389
11	0.0741	0.0515	3	15	0.0465
12	0.1852	0.0521	6	3.5	0.0577
13	0.1852	0.0521	6	3.5	0.0577
14	0.2222	0.0567	8	19.5	0.0871
15	0.2593	0.0547	7	2	0.0618

[a]*Criticality measure based on Eq. (5).*
[b]*Criticality measure based on Eq. (6).*
[c]*Holistic Indicator based on Eq. (7).*

Table 1.
Criticality evaluations based on four different metrics and computed holistic indicator.

In the last column of **Table 1**, we show the criticality rate for each node according to the new holistic indicator computed as in Eq. (7) considering $m=4$ metrics (i.e. the critical index, the Shapley Value, the node Degree and the Betweenness centrality). According to the procedure defined in Section 3, we have set the metric relevance as follows: $w_1 = 0.3$, $w_2 = 0.3$, $w_3 = 0.2$, and $w_4 = 0.2$ in order to emphasize the criticality metrics based on the concept of PWC.

The nodes color in **Figure 4** depends on the aggregated criticality values, according to the colormap. On the basis of this new indicator, the node 10 is the most critical node of the graph, in fact the deletion of this node strongly compromise the connectivity of the network and the creation of two disconnected partitions. Due to the same reason, a high criticality rate is also assigned to the nodes 4 and 3. Despite the node 14 is not essential for the connectivity, this node is characterized by a high node degree, in fact it is considered, according to the holistic indicator, as the fourth most critical node in the network.

Starting from the results obtained by computing the holistic indicator $\mathbf{r}^*$, we adopt a defensive strategy by defining a new attack cost $\hat{c}_i$, for each node, proportional to its holistic criticality rate as defined in Eq. (8). Notice that the defensive budget $B = \sum_{i=1}^{n} c_i = 15$.

The effectiveness of the proposed defensive strategy is proved by considering the global robustness index $\prod$, we remark that it came from the solution of Problem 1 and it is defined as the area under the Pareto front. As depicted in **Figure 5**, the new allocation of the defensive budget B is very effective to contrast an attacker especially with limited budget. In more details, in case of uniform defensive strategy (i.e. all the attack costs set to 1) the area under the Pareto front is equal to $\prod = 0.1229$, while the new budget allocation (Eq. (8)) based on the holistic indicator $\mathbf{r}^*$ improves the network robustness by increasing the area to $\prod = 0.1591$.

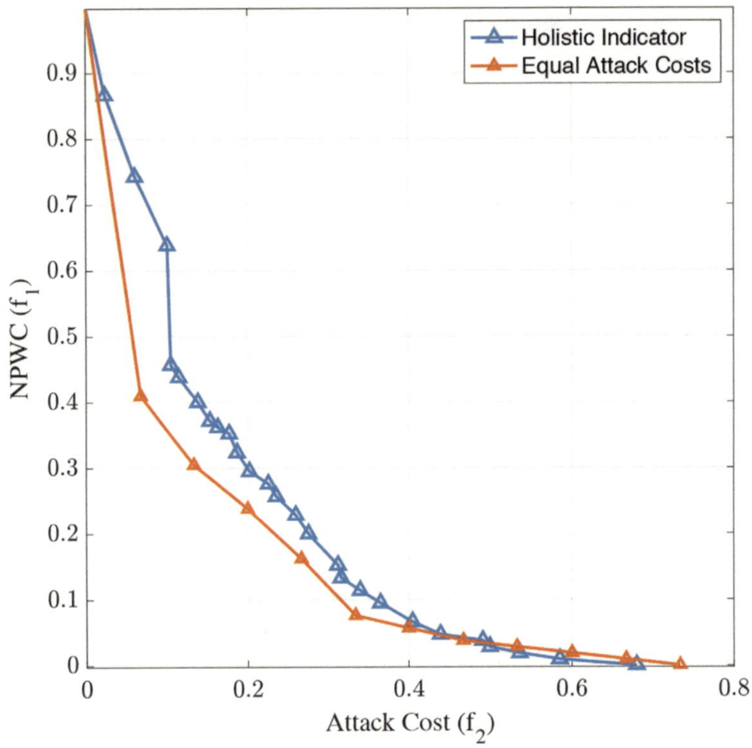

Figure 5.
Results of problem 1. Pareto fronts obtained by applying defensive strategies based on the holistic indicator (blue line), and uniform attack costs (red line).

6. Conclusions

In this chapter we provide a methodology for the definition of a defensive strategy via prioritizing the critical nodes of the network. Due to the complexity of a CI, the adoption of a unique metric for the identification of the node criticality is simplistic, to this end we propose a strategy, based on the AHP, able to merge multiple metrics which take into different aspects of the network. Moreover, the proposed aggregation procedure is applicable also in case of incomplete data. Among the multiple metrics applicable in the merging process, in this chapter we propose two metrics characterized by a focus on the network connectivity. In the one hand the critical index is computed on the basis of a multi objective optimization problem. Assuming an attacker perspective and knowing the topology of the network, the problems aims at identifying the nodes whose removal compromise the connectivity of the entire system. On the other hand, we propose the adoption of the Shapley value as a criticality evaluation by defining a cooperative game among the nodes of the network. Finally, we propose the definition of a defensive strategy that assigns to each node a removal cost proportional to the holistic indicator. Future improvement will be devoted to the inclusion of a final check able to include a final validation based on expert opinions. One of the possible validity check is based on the well-known face validity approach [19], it refers to the transparency or relevance of a test as it appears to test participants.

Acknowledgements

This work was supported by INAIL via the European Saf€ra project "Integrated Management of Safety and Security Synergies in Seveso Plants" (Saf€ra 4STER).

Author details

Luca Faramondi[1*], Giacomo Assenza[1], Gabriele Oliva[1], Ernesto Del Prete[2], Fabio Pera[2] and Roberto Setola[1]

1 Unit of Automatic Control, Department of Engineering, Università Campus Bio-Medico di Roma, Rome, Italy

2 National Institute for Insurance against Accidents at Work, Italy

*Address all correspondence to: l.faramondi@unicampus.it

IntechOpen

© 2021 The Author(s). Licensee IntechOpen. This chapter is distributed under the terms of the Creative Commons Attribution License (http://creativecommons.org/licenses/by/3.0), which permits unrestricted use, distribution, and reproduction in any medium, provided the original work is properly cited.

References

[1] Corsi, S., & Sabelli, C. (2004, June). General blackout in italy sunday september 28, 2003, h. 03: 28: 00. In IEEE Power Engineering Society General Meeting, 2004. (pp. 1691–1702). IEEE.

[2] Comfort, L. K., & Haase, T. W. (2006). Communication, coherence, and collective action: The impact of Hurricane Katrina on communications infrastructure. Public Works management & policy, 10(4), 328–343.

[3] Norio, O., Ye, T., Kajitani, Y., Shi, P., & Tatano, H. (2011). The 2011 eastern Japan great earthquake disaster: Overview and comments. International Journal of Disaster Risk Science, 2(1), 34–42.

[4] Weiss, J. (2016). Aurora generator test. Handbook of SCADA/Control Systems Security, 107.

[5] Assenza, G., Faramondi, L., Oliva, G., & Setola, R. (2020). Cyber threats for operational technologies. International Journal of System of Systems Engineering, 10(2), 128–142.

[6] Arulselvan, A., Commander, C. W., Elefteriadou, L., & Pardalos, P. M. (2009). Detecting critical nodes in sparse graphs. Computers & Operations Research, 36(7), 2193–2200.

[7] Arulselvan, A., Commander, C. W., Shylo, O., & Pardalos, P. M. (2011). Cardinality-constrained critical node detection problem. In Performance models and risk management in communications systems (pp. 79–91). Springer, New York, NY.

[8] Dinh, T. N., Xuan, Y., Thai, M. T., Park, E. K., & Znati, T. (2010, March). On approximation of new optimization methods for assessing network vulnerability. In 2010 Proceedings IEEE INFOCOM (pp. 1–9). IEEE.

[9] Faramondi, L., Oliva, G., Pascucci, F., Panzieri, S., & Setola, R. (2016, June). Critical node detection based on attacker preferences. In 2016 24th Mediterranean Conference on Control and Automation (MED) (pp. 773–778). IEEE.

[10] Faramondi, L., Setola, R., Panzieri, S., Pascucci, F., & Oliva, G. (2018). Finding critical nodes in infrastructure networks. International Journal of Critical Infrastructure Protection, 20, 3–15.

[11] Faramondi, L., Oliva, G., Panzieri, S., Pascucci, F., Schlueter, M., Munetomo, M., & Setola, R. (2018). Network structural vulnerability: a multiobjective attacker perspective. IEEE Transactions on Systems, Man, and Cybernetics: Systems, 49(10), 2036–2049.

[12] Faramondi, L., Oliva, G., & Setola, R. (2020). Multi-criteria node criticality assessment framework for critical infrastructure networks. International Journal of Critical Infrastructure Protection, 28, 100338.

[13] Oliva, G., Amideo, A. E., Starita, S., Setola, R., & Scaparra, M. P. (2019, September). Aggregating Centrality Rankings: A Novel Approach to Detect Critical Infrastructure Vulnerabilities. In International Conference on Critical Information Infrastructures Security (pp. 57–68). Springer, Cham.

[14] Saaty, T. L. (2008). Decision making with the analytic hierarchy process. International journal of services sciences, 1(1), 83–98.

[15] Faramondi, L., Oliva, G., & Setola, R. (2019, October). Network Defensive Strategy Definition Based on Node Criticality. In 2019 IEEE International Conference on Systems, Man and Cybernetics (SMC) (pp. 439–444). IEEE.

[16] Shapley, L. S., & Roth, A. E. (Eds.). (1988). "The Shapley value: essays in honor of Lloyd S. Shapley." *Cambridge University Press*.

[17] Saaty, T. L. (1977). A scaling method for priorities in hierarchical structures. Journal of mathematical psychology, 15 (3), 234–281.

[18] Biggs, N., Biggs, N. L., & Norman, B. (1993). Algebraic graph theory (Vol. 67). Cambridge university press.

[19] Nevo, B. (1985). Face validity revisited. Journal of Educational Measurement, 22(4), 287–293.

Chapter 7

Validation Strategy as a Part of the European Gas Network Protection

David Rehak, Martin Hromada, Ilias Gkotsis, Anna Gazi, Evita Agrafioti, Anastasia Chalkidou, Karolina Jurkiewicz, Fabio Bolletta and Clemente Fuggini

Abstract

The European gas network currently includes approximately 200,000 km high pressure transmission and distribution pipelines. The needs and requirements of this network are focused on risk-based security asset management, impacts and cascading effects of cyber-physical attacks on interdependent and interconnected European Gas grids. The European SecureGas project tackles these issues by implementing, updating, and incrementally improving extended components, which are contextualized, customized, deployed, demonstrated and validated in three business cases, according to scenarios defined by the end-users. Just validation is considered to be a key end activity, the essence of which is the evaluation of the proposed solution to determine whether it satisfies specified requirements. Therefore, the chapter deals with the validation strategy that can be implemented for the verification of these objectives and evaluation of technological based solutions which aim to strengthen the resilience of the European gas network.

Keywords: critical infrastructure, European gas network, validation, key performance indicators, resilience, protection

1. Introduction

The European gas network is an important and irreplaceable subsector of European Critical Infrastructure (ECI) [1]. The functioning of this network is constantly affected by threats with a direct but also cascading or synergistic effect [2]. These threats can be of various natures, e.g. meteorological, geological, process-technological, cascading, personnel, cyber or physical [3]. The impact of these threats can result in serious disruption or even failure of the regional parts of the gas network. For this reason, it is necessary to continuously improve the protection system of the European Gas Network, in particular through risk analysis and the consequent strengthening of the resilience through the identification and elimination of the identified weaknesses.

One of the main measures and means to achieve the enhancement of resilience, is through technological solutions, which should address the operational and technical needs of the infrastructure and requirements of the end user, i.e. infrastructure operator [4]. The chapter therefore deals with the validation strategy [5] that can be implemented for the verification of these objectives and the evaluation of technological based solutions which aim to strengthen the resilience of the

European gas network. The main objective of the proposed validation plan, as part of an overall evaluation process, is to study the acceptance of a designed security system aiming to promote resilience [6] of gas critical infrastructures (at strategic, tactical and operational level). For this purpose, it is necessary to collect qualitative information concerning some key criteria of the system which define its performance in the operations. The primary focus of the validation strategy is to assess the functionality and effectiveness of the proposed system. However, the intuitiveness of the individual components as well as the overall exploitation and operationalization potential of the developed solution, should also be evaluated.

The aforementioned validation plan has been developed and verified through continuous interaction with critical infrastructure (CI) operators within the SecureGas project [7]. The project aims to improve the resilience capabilities of the gas CI. The methodology uses a gas CI-contextualized Panarchy loop [8] reflecting a disaster life-cycle management process. The objective is to reduce foreseen risk, optimize the monetary investment, and reduce uncertainties. Providing the CI operators with a detailed validation methodological procedure to assess the added value of security solutions added to their infrastructure is of high value. Within the context of the SecureGas validation and evaluation, the following aspects that are addressed include: performance versus expectation, ease-of-use, understandability, reliability of operations, completeness and reliability of output, functionality, man–machine interface and efficiency. The criteria for validation, i.e. Key Performance Indicators (KPIs) [9], can be clustered into two categories: (1) general criteria that apply to the whole SecureGas system, and (2) specific criteria that apply to individual components of the system.

Such validation plan is fully transferable to other CI operators both of Gas and other sectors (e.g. power, telecommunication). With a slight adjustment of the identified KPIs, it can provide a valuable information on the applicability and usefulness of a security solution for risk mitigation, prevention and response purposes within a CI.

2. Validation, verification and evaluation

In order to understand the activities to be implemented from the validation point of view, definitions of the basic concepts used and are further analyzed below, presenting also several methodological approaches. Therefore, this section provides both a background analysis for validation-verification-evaluation processes and an adequate methodology.

The validation process involves the collection and evaluation of data, from the process design stage through commercial production phase, which establishes scientific evidence that a process meets a determined requirements. Process validation involves a series of activities taking place over the process. Regulatory authorities like European Medicines Agency and Food and Drug Administration have published guidelines relating to process validation [10]. The purpose of process validation is to ensure that varied inputs lead to consistent and high quality outputs. Process validation is an ongoing process that must be frequently adapted as manufacturing feedback is gathered. End-to-end validation of production processes is essential in determining product quality because quality cannot always be determined by a finished-product inspection. Process validation can be broken down into three steps: (1) process design, (2) process qualification, and (3) continued process verification.

The Guide to the Project Management Body of Knowledge (PMBOK guide), a standard adopted by the Institute of Electrical and Electronic Engineers, defines validation and verification as follows [5]:

- Validation: The assurance that a product, service, or system meets the needs of the customer and other identified stakeholders. It often involves acceptance and suitability with external customers. Contrast with verification.

- Verification: The evaluation of whether or not a product, service, or system complies with a regulation, requirement, specification, or imposed condition. It is often an internal process. Contrast with validation.

These terms generally apply broadly across industries and institutions. In addition, they may have very specific meanings and requirements for specific products, regulations, and industries. Some examples: Software [11], Food and drug, Health care [12], Greenhouse gas [13], Traffic and transport [14], Simulation models [15], ICT industry, Civil engineering [16], Economics, Accounting, Agriculture, Arms control.

In the context of the above, validation can generally be classified into five basic categories:

- Prospective validation comprises the missions conducted before new items are released to make sure the characteristics of the interests which are functioning properly and which meet safety standards [17]. Some examples could be legislative rules, guidelines or proposals [18–25].

- Retrospective validation is a process for items that are already in use in distribution or production. The validation is performed against the written specifications or predetermined expectations based upon their historical data/evidences that are documented/recorded. If any critical data is missing, then the work cannot be processed or can only be completed partially [10]. Retrospective validation is used for facilities, processes, and process controls in operation use that have not undergone a formally documented validation process. Validation of these facilities, processes, and process controls is possible by using historical data to provide the necessary documentary evidence that the process is doing what it is believed to do. Therefore, this type of validation is only acceptable for well-established processes and would be inappropriate where recent changes in the composition of product, operating processes, or equipment have occurred [26].

- Concurrent validation is used for establishing documented evidence that a facility and processes do what they purport to do, based on information generated during actual imputation of the process [26]. This approach involves monitoring of critical processing steps and end product testing of current production to show that the manufacturing process is in a state of control.

- Cross-validation is an approach by which the sets of scientific data generated using two or more methods are critically assessed [27].

- Re-validation is carried out for the item of interest that is dismissed, repaired, integrated/coupled, relocated, or after a specified time lapse. Examples of this category could be relicensing/renewing driver's license, recertifying an analytical balance that has been expired or relocated, and even revalidating professionals [28]. Re-validation may also be conducted when a change occurs during the courses of activities, such as scientific researches or phases of clinical trial transitions.

In contrast, evaluation is a systematic assessment of a subject's qualities, using criteria governed by a set of standards. Evaluation involves tests or studies conducted to investigate and determine the technical suitability of an equipment, material, product, process, or system for the intended objective. So evaluation can be formative that is taking place during the development of a concept or proposal, project or organization, with the intention of improving the value or effectiveness of the proposal, project, or organization. It can also be summative, drawing lessons from a completed action or project or an organization at a later point in time or circumstance. [29]

According to the way the evaluation is conducted we can distinguish the following types [30]:

- Internal evaluation, carried out by organizations, groups or stakeholders directly involved in the implementation of the project solution.

- External evaluation, carried out by specialists outside the development team, who are not employed within the organization responsible for the project under evaluation and who have no personal, financial or direct interest in the project.

Evaluation can be characterized as being either formative or summative. Broadly (and this is not a rule), formative evaluation looks at what leads to an intervention working (the process), whereas summative evaluation looks at the short-term to long-term outcomes of an intervention on the target group [31]:

- Formative evaluation takes place in the lead up to the project, as well as during the project, in order to improve the project design as it is being implemented (continual improvement). Formative evaluation often lends itself to qualitative methods of inquiry.

- Summative evaluation takes place during and following the project implementation, and is associated with more objective, quantitative methods.

Process evaluation is an inductive method of theory construction, whereby observation can lead to identifying strengths and weaknesses in program processes and recommending needed improvements [32]. For this purpose, qualitative methods are most often used, which are defined in the context of evaluation as research methods that emphasize depth of understanding, that attempt to tap the deeper meaning of human experience, and that intend to generate theoretically richer, observations which are not easily reduced to numbers [32]. The most used qualitative evaluation methods include [33]: content analysis, situational analysis, in-house surveys and interviewing.

Content analysis involves studying documents and communication artifacts, which might be texts of various formats, pictures, audio or video [34]. Quantitative content analysis highlights frequency counts and objective analysis of these coded frequencies [35]. Additionally, quantitative content analysis begins with a framed hypothesis with coding decided on before the analysis begins. These coding categories are strictly relevant to the researcher's hypothesis. Quantitative analysis also takes a deductive approach [36].

Situation analysis refers to a collection of methods that managers use to analyze an organization's internal and external environment to understand the organization's capabilities, customers, and business environment. The situation analysis consists of several methods of analysis: The 5Cs Analysis, SWOT analysis and Porter five forces analysis [37]. These analyses help understand the analytical processes by which managers understand themselves, their consumers, and the marketplaces in which they compete.

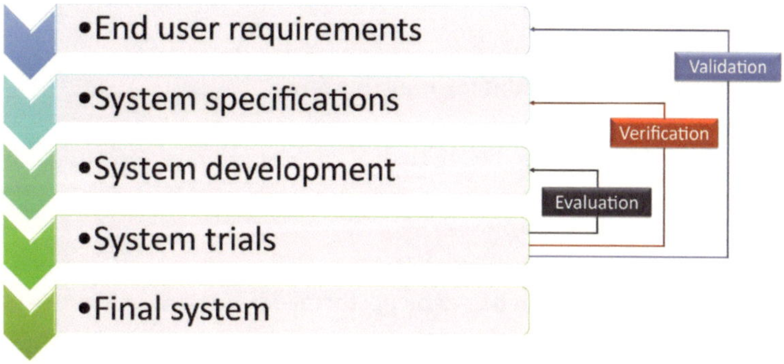

Figure 1.
Quality assurance framework.

SWOT analysis is a strategic planning technique used to help a person or organization identify strengths, weaknesses, opportunities, and threats related to business competition or project planning [38]. It is designed for use in the preliminary stages of decision-making processes and can be used as a tool for evaluation of the strategic position of an organization. It is intended to specify the objectives of the project and identify the internal and external factors that are favorable and unfavorable to achieving those objectives. Users of a SWOT analysis often ask and answer questions to generate meaningful information for each category to make the tool useful and identify their competitive advantage.

An interview is essentially a structured conversation where one participant asks questions, and the other provides answers. Interviews can range from Unstructured interview or free-wheeling and open-ended conversations in which there is no predetermined plan with prearranged questions [39], to highly structured conversations in which specific questions occur in a specified order [40].

Other commonly used tools and techniques for evaluation purposes [41] can include especially observation, survey questionnaires, case studies, analytical models, expert panel's consultation, cost–benefit analysis (CBA), and multi-criteria analysis (MCA).

Normally validation, verification and evaluation are performed in a row allowing to estimate the completeness and consistency of the system and examining its technical appropriateness, as depicted in **Figure 1**.

To sum up, verification and validation heavily rely on earlier phases of the project. Verification is a rather technical process in which the main question is whether the system works properly. The validation process covers not only the demonstrations but also earlier meetings and discussions in which the requirements are refined. As already mentioned, verification of developed tool/solution is the process of determining that the system is built according to its specifications. Validation is the process of determining that the system actually fulfills the purpose for which it was intended. Evaluation reflects the value and the acceptance of the system by the end users and its performance.

3. Concept of creating a validation plan

Following the analysis and presentation of validation, verification and evaluation processes, in this section, a holistic (including all those three processes) validation plan, will be analyzed. In principal, an effective validation and evaluation plan, needs to seek, as clear as possible, answers to the following issues:

1. What has to be evaluated?

2. Who is interested in the validation/evaluation?

3. What critical issues have to be tackled?

4. What has to be measured?

5. How validation/evaluation has to be performed?

6. Who is involved in the evaluation?

7. How results will be reported?

All these questions have been taken under consideration and are answered and described in detail as part of the SecureGas validation-evaluation methodological approach. In this four-step methodology (**Figure 2**), a set of business cases (BCs) is used to support the validation, verification and evaluation of SecureGas solution. Three BCs, addressing relevant issues for the gas sector (production, transport and distribution phase of the gas lifecycle, including different infrastructures for each phase) have been identified to ensure the delivery of solutions and services to the end-users. During the BCs implementation, tailor-made scenarios for the CIs will be used for demonstrations on actual sites. The technical components involved will be assessed quantitively (by measuring foreseen KPIs) and qualitatively (by using a set of questionnaires and interviews to the participants in the demonstrations).

3.1 Set the context

This kick-off step entails all the discussions and reviews with relevant stakeholders for the exact identification of the gaps and the existing capabilities. This step also sets the scope and the objectives of each BC for the SecureGas solution to provide differentiation from current practices and added value to the operational environment of a gas CI.

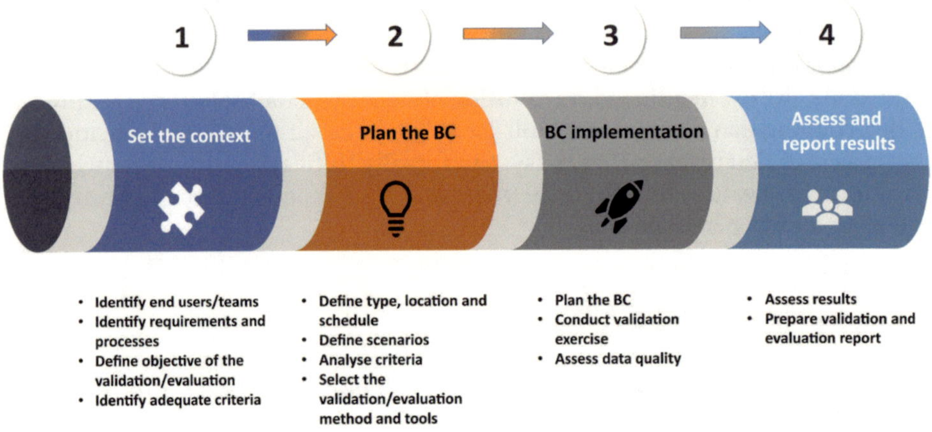

Figure 2.
SecureGas validation-evaluation methodology.

3.1.1 Identify end users/teams

Within SecureGas framework, the end-user team consists of the gas CI operators participating in the project (DEPA, EDAA, AMBER, ENI). Further to them, the SecureGas technical component providers are actively engaged and directly involved in all phases of the validation plan. External stakeholders have been identified and will be involved only in the BC implementation phase. They will participate and provide feedback for evaluation purposes. The stakeholders/actors participating in the pilot activities may vary among the different BCs however they belong to one of the following groups:

1. CI operators, managers and administrators, security liaison officers (also from interconnected, interdependent or similar CIs);

2. Emergency response authorities (police, fire brigade, civil protection, etc.);

3. National Authorities (CI regulatory authorities, ministries, etc.);

4. Security service providers;

5. Secondary/other security professionals and practitioners (e.g. policy makers, other EU research projects, etc.).

3.1.2 Identify requirements and processes

The SecureGas validation and evaluation process is an essential part of the project's development cycle. The development cycle is user-oriented, which means it relies on the perception, needs and responses by end users. Based on this development cycle, in SecureGas phase 1: "construct/develop", user requirements and specifications are identified leading to conceptual model (CM), concept of operations (ConOps) and high level reference architecture (HLRA). The CM, ConOps and HRLA will be implemented and demonstrated in phase 2: "demonstrate" and finally validated in phase 3: "validate & exploit".

Initial and crucial substeps to achieve an efficient planning and implementation of the BC are to:

1. Identify CI assets, threats, vulnerabilities, requirements, procedures, etc., in order to prepare the scenario including CI's specific security issues and addressing end users' actual needs.

2. Identify legacy systems and existing infrastructures, integration-data sharing, possible limitations, etc., and collaborate with the technical team to develop a SecureGas solution tuned to the project's BCs.

For the execution of these substeps, some may choose from a set of existing tools and frameworks, e.g. risk and vulnerability assessment and penetration testing (see Section 4).

3.1.3 Define the objective of the validation-evaluation process

The main objectives of the evaluation process will be to study the acceptance of the SecureGas system (at the strategic, tactical and operational levels), assess the performance of its components and the operational potential of the developed solution.

The beneficiaries of the validation and evaluation process are both technical component providers and CI operators. The technical providers will receive valuable feedback on technical development, components adaptation and implementation, system integration and cooperation with legacy systems, etc.. The CI operators will receive the performance assessment analysis of SecureGas solution, the extracted lessons, recommendations and conclusions, and all knowledge that can be transferred to their operations.

3.1.4 Identify adequate criteria

The criteria for validation can be clustered into two categories, further analyzed in Section 4:

- General criteria, that apply to the whole SecureGas system (cross-KPIs) and
- Specific criteria that apply to individual components of the system.

As such, the validation process will generate feedback during the pilot demonstrations on the following dimensions: functional, interface, security, operational, design, and implementation.

When it comes to the specific criteria, the SecureGas partners will make use of the lists of user (organizational, operational and regulatory) and technical (and standards-related) requirements defined, in order to determine whether the SecureGas system offers what it was designed to. As far as verification is concerned, the system specifications developed by technical partners will play the same role as user requirements in validation (see **Figure 1**). The evaluation process will also assess whether the SecureGas system complies with the technical requirements developed in Phase 1 of the project.

3.2 Plan the business case

This second part consists of a number of substeps that will lead in the realization of the BC implementation.

3.2.1 Type, location and schedule

In each SecureGas BC, an operational based demonstration will take place in the field (for the production, transport and distribution phases of gas lifecycle), aiming to simulate scenarios as realistically as possible in a controlled environment. This method of BC implementation will offer the advantage of real-time decisions and actions by the end-users and other participating actors, generating responses and leading to several consequences depending on the participants' actions and system performance. On top of that, regarding the strategic level of Gas lifecycle, a discussion-based approach will be followed, through the organization of a workshop/tabletop exercise, during which key personnel of the CI will have the chance to discuss scenarios that involve strategic threats and will assess policies, procedures, standard operating procedures and potential mitigation measures.

The locations may be related to the assets involved, the objectives and requirements of the validation, etc. Within SecureGas, the CI operators' sites in Greece, Lithuania and Italy have been selected and included in the scenarios based on the type of their installations.

Within the SecureGas project, project partners will customize, integrate and deploy the provided technical components into each BC. The deployment of the extended and integrated components in the BC will be tested through piloting activities for a period lasting almost one year period, with the last months focusing on the evaluations leading to an overall report based on the data and information collected.

3.2.2 Define scenarios

BCs are based on scenarios that correspond to a sequence of facts occurring in a specific space–time framework. Scenarios should be structured in a logical, readily accessible way to the pilot actors. Within SecureGas BCs, scenarios consist of events designed to guide the actors towards achieving the BC objectives. Six specific methodological substeps have been specified to define the scenarios:

Substep 1: Identification of normative, institutional and legislation frameworks.
Substep 2: Identification of end-user's infrastructures, assets and pilot site attributes.
Substep 3: Involved stakeholders and pilot actors.
Substep 4: Considered threats and risk.
Substep 5: Unfolding the scenario.
Substep 6: Deployment of the SecureGas solution.

3.2.3 Analyze criteria

The criteria used for the validation/evaluation of the SecureGas system and each component, consist of cross KPIs and specific KPIs (all linked with the end user requirements and technical specifications). In Section 4.1, these criteria will be discussed in detail.

3.2.4 Select validation/evaluation method and tools

In the framework of the validation plan, the methods and tools for the evaluation needs have been selected. Thus, the following substeps are executed for each BC:

1. Define what has to be measured for based on applicable KPIs.

2. Define how, through discussion-based workshop/tabletop exercise for the strategic level, and operations-based simulations/field pilots for the tactical/ operational level.

3. Define who are involved in the frame of the evaluation, sorted into three main groups as follows:

    - CI operators, security liaison officers, administrators and managers who can provide input based on an operational, policy and technical point of view, and evaluate the overall performance based on their experience.

    - First responders, who can provide input regarding the information sharing and community awareness during an incident.

    - Security practitioners and stakeholders, who, depending on their expertise, will provide information concerning the potential exploitation and use of the SecureGas solution. They may provide feedback on their willingness to use or adopt the system, other technical/operational comments, etc.

In order to achieve an effective evaluation outcome, the selection of the stakeholders, must be based on some requirements, such as the relevance to the scenario, adequate qualification, objectivity, previous experience.

4. Define the tools to be used to collect the results and feedback comprising:

    - KPIs and respective traceability matrices, for validation purposes, and
    - survey questionnaires, focus groups, interviews and brainstorming, for evaluation purposes.

5. Define how the results will be reported.

The results will be presented in suitable style and form, according to the reporting target audience and the selected tool. All reporting activities will be planned accordingly, paying attention to the most suitable communication means for the specific audience, in terms of content presentation, type of language, level of details and so on. For example, the elaboration of the questionnaires, the feedback from the interviews of the focus groups and the conclusions of the debriefing sessions (hot and cold washes) of BCs will be documented based on standardized feedback sheets which will be analyzed to improve the overall specification and development processes and their outcomes.

3.3 Business case implementation

The third part that will be followed in the validation plan, is that of that of the BC pilots execution, including both preparatory meetings and the actual field testing consisting of the following three substeps.

3.3.1 Plan the business case

1. End-users (internal and external) are identified specifically for each BC.
2. Identify the place and date and estimate the budget-plan logistics.
3. Send invitations, share information for the pilot with involved stakeholders.
4. Before the pilot, organize a training course, for the participants to have the opportunity to familiarize with the SecureGas solution.
5. The scenario (depending on the area of application) is presented to the end-users and its details are discussed.
6. All necessary adaptations, installations, integrations have been achieved and the system is ready to be used, demonstrated and evaluated.

3.3.2 Conduct validation exercise

Following the specific BC scenario storyline, the involved actors are guided and supported by the capabilities of the SecureGas system in order to respond to a security incident.

3.3.3 Assess data quality

Following the BCs pilots' implementation, the participants are asked to use the validation/evaluation tool/method (e.g. fill a specifically designed questionnaire, see Section 4). In some cases, interviews are held.

The assessment of results and feedback gathered leads to a holistic evaluation outcome, respective lessons identified and recommendations for further analysis.

3.4 Assess results

This last step of the methodology contains the analysis of the gathered evaluation results as well as an assessment of the SecureGas solution. The results of this step will be presented in the overall SecureGas evaluation and lessons identified report.

3.4.1 Assess results

The results assessment aims to collect valuable feedback from the end-users interactions during the pilots (via questionnaires, described in detail in Section 4.4), expressed opinions and comments through focus groups and end-session interviews. The purpose of this substep is to indicate among others whether the SecureGas solution is performing well, provides useful information, is easy to understand, reliable, ergonomic, efficient, etc.

3.4.2 Prepare validation and evaluation report

The final step in each BC pilot demonstration will summarize and present all the activities realized and the responses by involved actors' (both consortium partners and external experts). Based on these outcomes, an overall performance evaluation of the SecureGas solution will be reported, lessons, recommendations and conclusions will be extracted, and content for knowledge transfer will be structured.

4. Validation and evaluation tools

Within the SecureGas framework and specifically in the third phase of the project, that of validation and exploitation, several tools will be used in order to support the efficient implementation of the validation plan described in Section 3 above. These tools consists of: (a) an initial assessment tool, that will be used as a decision support tool to carry out a self-assessment to identify the level of intrusiveness and level of maturity of the CI, (b) the penetration testing tool/methodology for identifying vulnerabilities and assessing performance, (c) the KPIs that will be used as benchmarks to assess project's efficiency in reaching its key objectives and to evaluate the quality of the proposed technical solution, and finally (d) questionnaires and interviews as two main instruments for evaluation purposes.

4.1 Initial assessments

In the first step of the validation plan, the context is set as described in subSection 3.1. The validation plan follows the same approach as a pre-attack phase gathering as much information as possible on the target systems and planning the activities performed during the tests. Assessment frameworks such as [42, 43] can be used to identify the level of intrusiveness and level of maturity.

The substeps that are performed comprise:

1. Identify and prioritize assets: A list of identified assets indicating the importance of each one should be identified (e.g. software, hardware, data, interfaces, security governance, security controls and components, etc.).

2. Identify threats: A threat is anything that could exploit a vulnerability to breach security and cause harm to a CI. General threat categories are: physical adversarial threats and acts of terrorism, political/geopolitical/social threats, natural hazards, technological and accidental hazards, indirect threats and cyber threats.

3. Identify Vulnerabilities: Identify a list of known vulnerabilities of all the asset list and analyze the impact on the system/infrastructure if these are not correctly treated and mitigated The impact on the system shall be treated in terms of e.g. economy, reputation, and security for people

4. Analyze measures: Analyze the measures that are either in place or in the planning stage to minimize or eliminate the probability that a threat will exploit a vulnerability in the system

5. Determine the likelihood of an incident: The possibility of an incident to be an exploited vulnerability should be quantified, based on historical/statistical data, user experience and knowledge or any other sources available (e.g. studies, estimations/information that authorities are producing, etc.).

6. Assess the impact a threat could have, including factors such as the mission, the criticality and the sensitivity of the system and its data

7. Prioritize the security risk: For each threat/vulnerability pair, determine the level of risk for the system/infrastructure, based on the likelihood and the impact of the threat, and the adequacy of the existing or planned system/infrastructure security controls for eliminating or reducing the risk

8. Recommend Controls: Using the risk level from the previous step, determine the actions that the senior management of the CI and other personnel that hold key positions, must take to mitigate the risk to an accepted residual risk level.

9. Document the results to support management in making appropriate decisions on budget, policies, procedures, and so on.

4.2 Penetration testing

Following the above assessment, another process that can be used as a tool for identifying vulnerabilities and assessing performance is Penetration Testing (PT). PT is a security testing process in which experts execute real but yet controlled attacks on systems and services to identify methods for circumventing the security features of an application, system, or network [44].

PT methodologies divide the process into four generic phases:

1. A planning phase, focuses on gathering available information on the target systems, as well as on potential methods of attacks, management approval and setting the groundwork for setting up attack strategies and attack scenarios.;

2. A discovery phase, which is broken down into two parts: information gathering and scanning, and vulnerability analysis;

3. An attack Phase, where the tester put in place the knowledge acquired in the previous phase. This phase contains the following substeps: (a) Gaining access, (b) escalating privileges, (c) System browsing, and (d) Install additional tools;

4. A reporting phase, where experts evaluate findings and propose corrective actions.

4.3 Key performance indicators

KPIs typically enable the realization of technical systems towards tangible goals while serving as a benchmark for internal quality assurance. Indeed, KPIs are deemed as a measurable way to assess project's efficiency in reaching its key objectives and to evaluate the quality of the proposed technical solution(s). Through well-defined KPIs, the main areas to be tested, measured and validated during the piloting activities are established.

The SecureGas KPIs were defined in the early stage of the project so that they guide its targeted implementation. Preliminary activities, regarding user and system requirements identification as well as the CONOPS and HLRA definition, have already been completed providing valuable input to the KPIs definition task.

For the purposes of the SecureGas project, the KPIs were classified along two main indicator types:

a. SecureGas component KPIs, which reflect the key characteristics and functionalities offered by each SecureGas component and are applied for their performance evaluation;

b. SecureGas Cross-KPIs, which reflect the key functionalities and the expected quality of the entire SecureGas solution.

Both the SecureGas component KPIs and the SecureGas Cross-KPIs establish the validation criteria to be measured during SecureGas pilot demonstrations. Although both KPI categories are equally important for the evaluation of objectives' fulfillment, this section emphasizes on the KPIs defined for the integrated SecureGas system (i.e. SecureGas Cross-KPIs).

The methodology adopted for the definition of the KPIs was built on a bottom-up rationale. The SecureGas component KPIs (low level KPIs) were initially defined. Then, drawing on that information, the SecureGas Cross-KPIs (high level KPIs) were derived. The procedural pathway followed for the identification of KPIs is depicted in **Figure 3**.

Considering that KPIs depend on the end-users and stakeholders interested in the SecureGas system, the first step of the adopted methodology regarded their active engagement in the KPIs definition activities. This initiative had already started taking place through the definition of the user requirements (i.e. end-users needs and expectations from an integrated security system (such as the SecureGas system), as well as through dedicated stakeholders' workshops organized for the user requirements validation. The user requirements together with their external validation results shed light to those characteristics of the system that are deemed important by the end-users. In addition, information on the KPIs already applied by the end-users to assess the performance of their gas network daily operations allowed consortium partners to draft broad areas in which evaluations are

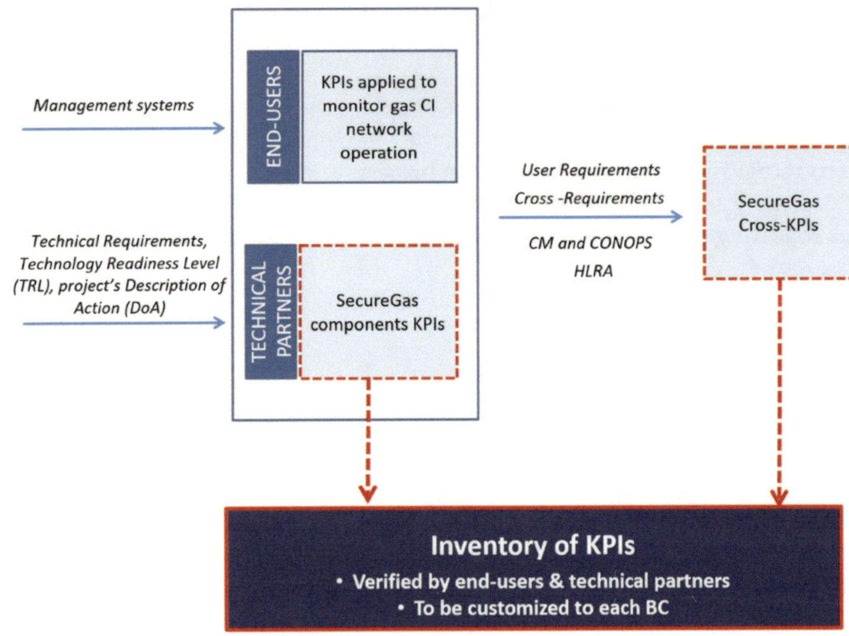

Figure 3.
KPIs definition pathway.

performed. This information also enabled the consortium to examine how the SecureGas solution could contribute and add value to the resilience of end-users' infrastructure.

In parallel, drawing on the already defined technical requirements of the SecureGas components, consortium technical partners defined the key capabilities, characteristics and functionalities offered by every technical subsystem. The so-called SecureGas component KPIs enable components' development and implementation.

The next step regarded the definition of the SecureGas Cross-KPIs which reflect the most important features and characteristics offered by the entire (i.e. all subsystems integrated into one system) SecureGas solution. The end-users KPIs, the SecureGas component KPIs and the already defined SecureGas system specifications (Cross-Requirements), provided the baseline for the extraction of a list of eleven SecureGas Cross-KPIs (**Table 1**) that are key to performance success.

As presented in **Table 1**, the SecureGas Cross-KPIs were classified into specific Fields that outline the general domain categories where the impacts are going to exert their effect. Those Fields are as follows:

- Reliability, i.e. the capability of the system to function in a correct manner within the given timeframe. This includes high accuracy of alert localization, avoidance of any delays in data provision, and a low rate of false alerts or errors.

- Autonomy, i.e. the level of independence of the system. An autonomous system is capable to operate (detect and process incidents) without human supervision (human in the loop only when deemed necessary).

- Interoperability, i.e. the ability of the system to work with new products (i.e. sensors or sub-systems) without special configurations.

Field	Indicator	Description	Metric	Target value
Reliability	False alert rate	Percentage of false alerts (both positive and negative) raised by the SecureGas system.	% (False alerts / Total alerts)	< 5%
	Cross correlation	Percentage of cross correlated alerts raised by the SecureGas system.	% (Cross correlated alerts / Total alerts)	> 50%
	Latency	Time elapsed between the moment an incident occurs and the moment the alert is displayed in the operational picture.	Time (sec)	< 10 sec
	Mean time to notify	Time needed for the operator to create an incident notification and send it to competent authorities/ stakeholders (escalation of incident).	Time (min)	< 3 min
Autonomy	Threat categories addressed	Number of different threats categories addressed by the SecureGas system (Threat categories: cyber, physical, cyber-physical, physical-cyber)	Number	4
	Automatic detection of threats	Number of different threat types automatically detected by the system. (Threat types: Intrusion detection, Third-Party Interference, Leak, Landslide hazard, Cyber)	Number	≥5
	Automatic decision-support	Percentage of alerts automatically linked to recommendations on crisis management and mitigation actions	% (Alerts with decision support / Total alerts)	≥ 80
Interoperability	Transparent integration of users' legacy systems	Number of users' legacy systems that can be easily and transparently integrated into the SecureGas system.	Number	≥1
Usability	Multilingual interface	Number of different languages in which the SecureGas user interface will be available	Number	4 (English, Italian, Greek, Lithuanian)
Resilience	Self-testing capabilities (system health check)	Percentage of components/ sensors that provide information to the operator - through dedicated alerts - about their status (not functioning and/or no communication)	%	90–95%
	Accuracy degradation percentage of a measurement value	The maximum decrease of accuracy (due to concept drift), before the model is retrained to adapt to background changes	%	20%

Table 1.
SecureGas cross-KPIs.

- Usability, i.e. is a set of attributes covering the effort needed for using a solution, and on the individual assessment of the use of the solution, by a stated or implied set of users.

- Resilience, i.e. is the ability of the SecureGas system to adapt from a disruption. This means that the system is able to identify potentially disruptive events and adapt to the evolving circumstances.

Each of the aforementioned Fields was linked to a set of Indicators, each one being assigned a Description, Metric and Target Value.

Following the main principles of the SecureGas project, the SecureGas Cross-KPIs aimed and achieved to addresses all the Risk and Resilience phases. Those phases reflect the activities that need to be conducted before, during and after disruptive events, as part of a comprehensive risk and resilience management procedure. The Risk and Resilience phases are as follows: Prepare, Detect, Prevent, Absorb, Respond, Recover, Learn and Adapt. The ultimate goal of developing Cross-KPIs for all those phases was to showcase how the core functionalities and performance indicators of the SecureGas system can add value to the enhancement of the resilience of gas critical infrastructure networks. **Figure 4** presents the Risk and Resilience phases that are affected by each SecureGas Cross-KPIs. Some of the Cross-KPIs are linked to one phase, some others to more, while the Cross-KPI "Multilingual Interface" is related to all the seven Risk and Resilience phases, since the enhancement of the usability parameters of a system has the potential to affect the entire security and resilience status of a CI network.

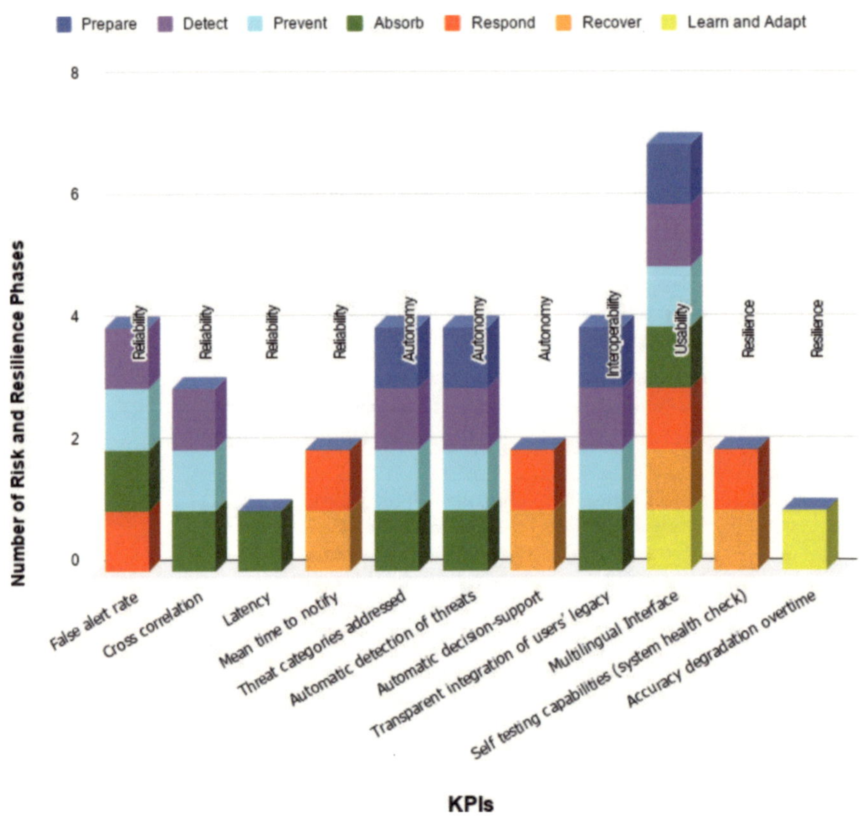

Figure 4.
Risk and resilience phases affected by each SecureGas cross KPI.

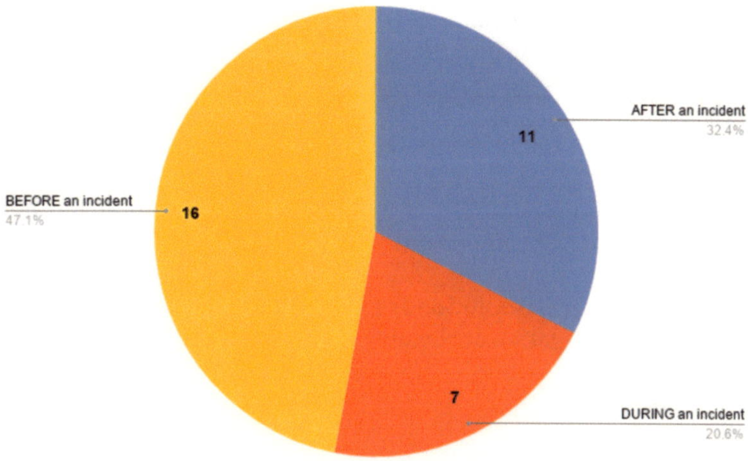

Figure 5.
KPIs distribution to the activities taking place before, during and after an incident.

Figure 5 shows the KPIs distribution to the activities taking place before, during and after incidents. In general, the SecureGas Cross-KPIs are mostly linked to the activities/phases taking place before the occurrence of an incident (prepare, detect, prevent) (approx. 47.1% of KPIs), although the SecureGas system do have performance parameters that are related to the post incident activities (response, recover, learn and adapt) (approx. 32.4%).

4.4 Questionnaires and interviews

Within the context of the evaluation of SecureGas components and solution, two main instruments will be used: questionnaires and interviews.

Regarding the first one, two types of questionnaires will be used for the evaluation purposes, one more generic that can be distributed to all participants (during testing, demonstrations, workshops) and one more specific, that would be filled by targeted participants within the audience, as further described below:

1. Questionnaire 1 (generic): This will be addressed to all participants of the BC demonstrations and is based on the System Usability Scale (SUS), developed by John Brooke in 1986 [45]. The questionnaire 1 provides a "quick and dirty" though reliable tool for measuring the usability of tested systems. SUS consists of a 10-item questionnaire with five response options for respondents; from strongly agree to strongly disagree. This allows to gather evaluation feedback concerning a wide variety of products, systems and services, including hardware, software, mobile devices, websites and applications. SUS has become an industry standard, with references in several articles and publications.

2. Questionnaire 2 (specific): The second questionnaire aims to extract end-users' assessed indicators on the basis of intuitiveness, usability, performance, etc. of the proposed solution. The end-users are going to fill-in this specific questionnaire after they have experienced the capabilities and the use of the system during the BC demonstration. This questionnaire is divided in seven main sections (i.e. general information, ease of installation, facilitation of user learning, data requirements, integrity, usability, usefulness), each one aimed at examining a different aspect of the end-users' view on the SecureGas components.

Regarding the second instrument for evaluation, indicative topics that may be used for discussion during the interviews comprise:

1. Experience and comments on the parallel processing, dataflow and cooperating applications within the SecureGas system.

2. Integration and interoperability of components, input/output and automatic/manual procedures for components.

3. Evaluation of SecureGas solution as a whole for the identification, detection, assessment and mitigation of threats and risk.

5. Conclusions

The validation framework is a key activity of every project, which broadly includes the validation of the proposed solution to determine whether it satisfies specified requirements, the verification of the system specifications, and the evaluation of the developed solution, all further analyzed as processes in Section 2. In the framework of the SecureGas project, the developed solution is a set of technological components and practical tools which aim to strengthen the resilience of the European gas network.

The envisaged validation framework (Section 3) mainly includes two types of assessment (Section 4): (a) Quantitative assessment, using a series of KPIs to validate components and the solution as a whole, (b) Qualitative assessment, based upon a dedicated questionnaire and interview, to get feedback from participants in the BCs implementation.

The methodological procedure, described in Section 3 of this chapter, is of no doubt necessary for any technological team providing a solution in order to identify potential gaps and updates needed. Furthermore, it is also valuable for end-users, in order to recognize the suitability of the proposed solution based on their requirements and specific security issues and appreciate the added value offered. Such validation framework is applicable, at least as a concept, to all projects offering technological solutions towards CI operators (or other type of end users) and can be adapted and tailor made to each case, leading to valuable feedback. On the other hand, the proposed methodology may need some adjustments, in order to cover the needs of an end-user that would like to assess and validate a process or a procedure that may have already in hand or is proposed (e.g. KPIs redefinition, questionnaires restructuring, etc.).

The next steps of this research contain the implementation of the BCs, based on this validation plan, and the documentation of the results of each BC, consolidating them into an overall validation and performance evaluation, which may lead to lessons identified, best practices and recommendations for the interested stakeholders.

Acknowledgements

The chapter was supported by the European Commission [Project SecureGas, GA No 833017] and the Ministry of the Interior of the Czech Republic [Project CIRFI 2019, GA No VI20192022151].

Conflict of interest

The authors declare no conflict of interest.

Author details

David Rehak[1]*, Martin Hromada[2], Ilias Gkotsis[3], Anna Gazi[3], Evita Agrafioti[4], Anastasia Chalkidou[4], Karolina Jurkiewicz[5], Fabio Bolletta[6] and Clemente Fuggini[6]

1 VSB – Technical University of Ostrava, Faculty of Safety Engineering, Ostrava, Czech Republic

2 Technology Platform Energy Security, Prague, Czech Republic

3 KEMEA - Center for Security Studies, Athens, Greece

4 GAP Analysis S.A., Athens, Greece

5 APRE – Agenzia per la Promozione della Ricerca Europea, Rome, Italy

6 Rina Consulting S.p.A., Genoa, Italy

*Address all correspondence to: david.rehak@vsb.cz

IntechOpen

© 2020 The Author(s). Licensee IntechOpen. This chapter is distributed under the terms of the Creative Commons Attribution License (http://creativecommons.org/licenses/by/3.0), which permits unrestricted use, distribution, and reproduction in any medium, provided the original work is properly cited.

References

[1] Council Directive 2008/114/EC of 8 December 2008 on the identification and designation of European critical infrastructures and the assessment of the need to improve their protection. Brussels: Council of the European Union.

[2] Rehak D, Senovsky P, Hromada M, Lovecek T, Novotny P. Cascading Impact Assessment in a Critical Infrastructure System. International Journal of Critical Infrastructure Protection. 2018:22:125-138. DOI: 10.1016/j.ijcip.2018.06.004

[3] Rehak D, Senovsky P, Hromada M, Lovecek T. Complex Approach to Assessing Resilience of Critical Infrastructure Elements. International Journal of Critical Infrastructure Protection. 2019:25:125-138. DOI: 10.1016/j.ijcip.2019.03.003

[4] Sullivant J. Strategies for Protecting National Critical Infrastructure Assets: A Focus on Problem-Solving. Hoboken, NJ: Wiley; 2007.

[5] IEEE Draft Guide: Adoption of the Project Management Institute (PMI) Standard: A Guide to the Project Management Body of Knowledge (PMBOK Guide)-2008 (4th edition). Piscataway, NJ: Institute of Electrical and Electronics Engineers; 2011.

[6] NIAC (National Infrastructure Advisory Council). Critical Infrastructure Resilience Final Report and Recommendations. Washington, DC: U.S. Department of Homeland Security; 2009.

[7] SecureGas project. Securing the European Gas Network [Internet]. 2020. Available from: https://www.securegas-project.eu/ [Accessed: 2020-09-18]

[8] Allen CR, Angeler DG, Garmestani AS, Gurdenson LH, Holling CS. Panarchy: Theory and Application. Ecosystems. 2014:17:578-589. DOI: 10.1007/s10021-013-9744-2

[9] Badawy M, El-Aziz AA, Idress AM, Hefny H, Hossam S. A Survey on Exploring Key Performance Indicators. Future Computing and Informatics Journal. 2016:1(1-2):47-52. DOI: 10.1016/j.fcij.2016.04.001

[10] Food and Drug Administration. Guidance for Industry Process Validation: General Principles and Practices [Internet]. 2011. Available from: https://www.fda.gov/files/drugs/published/Process-Validation--General-Principles-and-Practices.pdf [Accessed: 2020-08-04]

[11] Pham H. Software Reliability. Hoboken, NJ: John Wiley & Sons; 1999.

[12] Haggas R. Validation of electronic issue on the lth blood bank telepath system [Internet]. 2007. Available from: https://web.archive.org/web/20071012043133/http://www.transfusionguidelines.org.uk/docs/pdfs/oig_tools_qa_bb_e-issue_validation.pdf [Accessed: 2020-08-09]

[13] ISO 14064-1:2018. Greenhouse Gas Emissions and Removals Quantification and Reporting.

[14] Apeltauer T, Macur J, Holcner P, Radimsky M. Validation of microscopic traffic models based on gps precise measurement of vehicle dynamics. Promet – Traffic&Transportation. 2013:25(2):157-167. DOI: 10.7307/ptt.v25i2.1293

[15] Sargent RG. Verification and validation of simulation models. In: Jain S, Creasey RR, Himmelspach J, White KP, Fu MC, editors. Proceedings of the 2011 Winter Simulation Conference (WSC'11); December 2011; Phoenix, AZ: Winter Simulation Conference; 2011. p. 183-198.

[16] De Graaf RS, Vromen RM, Boes J. Applying systems engineering in the civil engineering industry: an analysis of systems engineering projects of a Dutch water board. Civil Engineering and Environmental Systems. 2017:34(2):144-161. DOI: 10.1080/10286608.2017.1362399

[17] Food and Drug Administration. Guideline on general principles of process validation [Internet]. 1987. Available from: https://web.archive.org/web/20090606085627/https://www.fda.gov/Drugs/GuidanceCompliance RegulatoryInformation/Guidances/ucm124720.htm [Accessed: 2020-08-16]

[18] Quinn J, McDermott D, Stiell I, Kohn M, Wells G. Prospective Validation of the San Francisco Syncope Rule to Predict Patients With Serious Outcomes. Annals of Emergency Medicine. 2006:47(5):448-454. DOI: 10.1016/j.annemergmed.2005.11.019. PMID 16631985

[19] Sangiovanni A, Manini M, Iavarone M, Fraquelli M, Forzenigo L, Romeo R, Ronchi G, Colombo M. Prospective validation of AASLD guidelines for the early diagnosis of epatocellular carcinoma in cirrhotic patients. Digestive and Liver Disease. 2007:40(5):A22-A23. DOI: 10.1016/j.dld.2007.12.064

[20] Germing U, Strupp C, Kuendgen A, Isa S, Knipp S, Hildebrandt B, Giagounidis A, Aul C, Gattermann N, Haas R. Prospective validation of the WHO proposals for the classification of myelodysplastic syndromes. Haematologica. 2006:91(12):1596-1604.

[21] Sciolla R, Melis F. Rapid Identification of High-Risk Transient Ischemic Attacks: Prospective Validation of the ABCD Score. American Heart Association. 2008:39(2):297-302. DOI: 10.1161/STROKEAHA.107.496612

[22] Pfisterer M, Bertel O, Bonetti PO, Brunner-La Rocca HP, Eberli FR, Erne P, Galatius S, Hornig B, Kiowski W, Pachinger O, Pedrazzini G, Rickli H, De Servi S, Kaiser Ch. Drug-eluting or bare-metal stents forlarge coronary vessel stenting? The BASKET-PROVE (PROspective Validation Examination) trial: Study protocol and design. American Heart Journal. 2008:115(4):609-614. DOI: 10.1016/j.ahj.2007.11.011

[23] Van Geest-Daalderop JHH, Hutten BA, Péquériaux NCV, Levi M, Sturk A. Improvement in the regulation of the vitamin K antagonist acenocoumarol after a standard initial dose regimen: prospective validation of a prescription model. Journal of Thrombosis and Thrombolysis. 2008:27(2):207-214. DOI: 10.1007/s11239-008-0203-4

[24] Ames D, Keogh AM, Adams J, Harrigan S, Allen N. Prospective validation of the EBAS-DEP – A short sensitive screening instrument for depression in the physically ill elderly. European Psychiatry. 1996:11(4):361s. DOI: 10.1016/0924-9338(96)89148-6

[25] Kidwell ChS, Starkman S, Eckstein M, Weems K, Saver JL. Identifying Stroke in the Field: Prospective Validation of the Los Angeles Prehospital Stroke Screen (LAPSS). American Heart Association. 2000:31(1):71-76. DOI: 10.1161/01.str.31.1.71

[26] Kneat Solutions. The Four Types of Process Validation [Internet]. 2017. Available from: http://blog.kneat.com/the-four-types-of-process-validation [Accessed: 2020-08-25]

[27] Food and Drug Administration. Bioanalytical Method Validation Guidance for Industry [Internet]. 2018. Available from: https://www.fda.gov/regulatory-information/search-fda-guidance-documents/bioanalytical-method-validation-guidance-industry [Accessed: 2020-08-28]

[28] Merkur S, Mossialos E, Long M, McKee M. 2008. Physician revalidation in Europe. Clinical Medicine Journal. 2008:8(4):371-376. DOI: 10.7861/clinmedicine.8-4-371

[29] Scriven M. The methodology of evaluation. Lafayette, IN: Purdue University; 1966.

[30] Volkov BB, Baron ME. 2011. Issues in internal evaluation: Implications for practice, training, and research. New Directions for Evaluation. 2011:132:101-111. DOI: 10.1002/ev.399

[31] Owen JM, Rogers PJ. Program Evaluation: Forms and Approaches. Thousand Oaks, CA: Sage Publications; 1999.

[32] Rubin A, Babbie E. Research methods for social work. 4th ed. Belmont, CA: Wadworth/Thomas Learning; 2001.

[33] Bess G, King M, LeMaster PL. Process evaluation: How it works. American Indian and Alaska Native Mental Health Research. 2004:11(2):109-120.

[34] Bryman A. Business research methods. 3rd edit. Cambridge: Oxford University Press; 2011.

[35] Kracauer S. The Challenge of Qualitative Content Analysis. Public Opinion Quarterly. 1952:16(4):631-642. DOI: 10.1086/266427

[36] White MD, Marsh EE. Content Analysis: A Flexible Methodology. Library Trends. 2006:55(1):22-45. DOI: 10.1353/lib.2006.0053

[37] Steenburgh T, Avery J. Marketing Analysis Toolkit: Situation Analysis. Boston, MA: Harvard Business School; 2010.

[38] Humphrey A. SWOT Analysis for Management Consulting. Menlo Park, CA: SRI International; 2005.

[39] Yale JR. Frontier Thinking in Guidance. Chicago, IL: Science Research Associates; 1945.

[40] Kvale S, Brinkman S. Interviews: Learning the Craft of Qualitative Research Interviewing. 2nd ed. Thousand Oaks, CA: Sage; 2009.

[41] Morra-Imas LG, Rist RC. The road to results: designing and conducting effective development evaluations. Washington, DC: The World Bank; 2009.

[42] ISO/IEC 27001:2013. Information security management.

[43] ISO/IEC 27005:2018. Information security risk management.

[44] NIST 800-115:2008. Technical Guide to Information Security Testing and Assessment.

[45] UsabiliTEST. System Usability Scale (SUS) Plus [Internet]. 2020. Available from: https://www.usabilitest.com/system-usability-scale [Accessed: 2020-09-12]

Chapter 8

Defects Assessment in Subsea Pipelines by Risk Criteria

Anatoly Lepikhin, Victor Leschenko and Nikolay Makhutov

Abstract

Subsea inter-field pipelines are an important element of offshore oil and gas infrastructure. Leakage or fracture of these pipelines is associated with the risk of large economic and environmental losses. One of the main sources of pipeline fracture is pipe defects. The presented section discusses the methodological aspects of assessing the hazard of defects of subsea inter-field pipelines by risk criteria of accidents. A conceptual approach of defects hazard assessing by risk criteria has been formulated, based on analysis the requirement of modern standards. The risk is defined as the probability of negative consequences, the scale of which is determined by the hazard class of pipeline accidents. The probability and scale of accidents are linked by a risk matrix. A method for a three-level assessment of the suitability of a pipeline for operation after in-line inspection has been developed. The method allows assessing the hazard of the most typical defects in subsea pipelines, such as metal loss, metal delamination, cracks and crack-like defects. The allowable defect sizes are determined for the given risk criteria using partial safety factors. The novelty of the methodology lies in the substantiation of safety factors according to risk criteria corresponding to a given class of damage and loss. A scheme for making decisions on the admissibility of defects by risk criteria has been developed. An example of hazard assessment of defects in subsea pipelines is presented.

Keywords: subsea inter-field pipelines, defect, fracture, criterion, risk, calculation

1. Introduction

Subsea inter-field pipelines are an important element of the offshore oil and gas condensate field infrastructure. Leakage or breakdown of these pipelines is associated with the risk of large economic and environmental damage. To ensure the safe operation of pipelines, systematic non-destructive testing is carried out using in-line diagnostics. Production, construction and operational defects of pipes are often by found the diagnostics. The presence of defects in the pipes requires solving the problem of classification and risk assessment of defects. In the classical setting, this problem is solved on the basis of the norms allowable defect sizes [1, 2]. The unacceptable defects are subject to mandatory repair or elimination. This approach is irrationality and has been repeatedly discussed and criticized from various points of view [3–5]. Classifications of defects on the basis of calculation their hazard, taking into account the peculiarities of the operating conditions are more reasonable. At the moment such calculation base on using a number of methods [6–10]. However, these methods also have a number of disadvantages. The most significant disadvantage is that they are based on a deterministic concept of ensuring strength, with deterministic defects sizes, loads values and characteristics of mechanical properties of pipe metal. In real conditions, random variations and statistical scattering of calculated variables always occur, which violate the uniqueness of the estimates of the hazard of defects. Taking this into account, the methods assessment of defects hazard based on the normative approach and deterministic strength calculations can be considered justified during the construction or reconstruction of pipelines. But they are irrational at the stage of pipeline operation, when deviations from design solutions, specified technological modes, environmental conditions and other factors affecting the performance arise. In such conditions, some of the permissible defects can be dangerous, and vice versa, pipelines with defects that are unacceptable according to the norms can be (and often turn out to be) workable.

Probabilistic risk analysis methods develop to assess the operability of structures with defects [11]. In these methods, the defect hazard is determined by the level of risk of pipeline destruction. This ensures, on the one hand, taking into account the probabilities of violation of the strength conditions in the presence of defect, on the other hand, taking into account the severity of the consequences of accidents. This article discusses the methodological aspects of assessing the defects hazard in subsea inter-field pipelines according by the criteria of the risks of destruction. Risk is understood as the probability of losses from leakage or pipeline failure, caused by the considered defects. This formulation of the problem differs significantly from the above-mentioned traditional approaches to assessing the defects hazard, based on strength calculations.

2. Accident analysis of subsea pipelines

The safety of operation subsea pipelines is ensured by using modern methods of design, manufacture, operation and maintenance, regulated by the rules and regulations. Nevertheless, the practice of operating pipelines is accompanied by cases of fracture with negative consequences. Currently, a large amount of statistical data has been accumulated on accidents of onshore and subsea pipelines. Statistical data on emergency conditions for subsea pipelines qualitatively and quantitatively differ from statistics on emergency and underground pipelines due to differences in operating conditions and modes. Therefore, the statistical data for surface and underground pipelines can only be taken into account for qualitative comparisons.

Accident rate statistics for subsea pipelines are mainly presented for the water areas and continental shelf of the North Sea (PARLOC database) and the Gulf of Mexico (DOT database) [12]. These data cover the period from 1984 to the present, with operating experience over 480 thousand km × year (PARLOC base) and over 650 thousand km × year (DOT base). According to PARLOC data the average failure rate is 8.79×10^{-5} 1/km × year, and according to DOT data is 3.51×10^{-4} 1/km × year. For comparison, according to the UKOPA database (Great Britain), which includes statistics on underground pipelines, with experience over 700 thousand km × year, the average failure rate is 4.86×10^{-5} 1/km × year. According to the EGIG database (European Union), which also includes data on the accident rate of underground pipelines, with experience over 3150 thousand km × year, the average failure rate is 3.70×10^{-4} 1/km × year. According to statistics, the main reasons for failure of subsea pipelines are:

- mechanical damage (hooking with anchors and trawls, falling heavy objects);
- corrosion and aging processes;

- construction and pipe metal defects;
- natural impacts (landslides, earthquakes, underwater currents, etc.).

At the same time, the average failure rate due to corrosion is in the range $(1.16 \times 10^{-6}–4.21 \times 10^{-4})$ 1/km × year (PARLOC data) and in the range $(1.01 \times 10^{-5}–7.10 \times 10^{-5})$ 1/km × year (DOT data). The average failure rate due to external influences is in the ranges: DOT is $(5.52 \times 10^{-6}–1.3 \times 10–^{4})$ 1/km × year; PARLOC is $(1.53 \times 10^{-5}–9.46 \times 10^{-5})$ 1/km × year.

According to the data [13] for the period 1970–2009 years 6183 accidents of subsea pipelines occurred in the world. The main number of accidents was recorded in the North Sea (3505) and the Gulf of Mexico (1658). In the Mediterranean Sea, the number of accidents was 45, in the Black and Caspian Seas – 29 accidents. At the same time, up to 41% of accidents occur due to external reasons, and up to 47% of accidents due to pipe defects. According to [14], 95 accidents occurred on the continental shelf of Great Britain in 2012–2013 years, of which 49 accidents occurred due to mechanical reasons (defects, fatigue, corrosion, erosion).

Of particular interest are assessments of damage from pipeline accidents. Unfortunately, such data are rarely published. In the above-mentioned work [12], it is noted that the total damage from 125 accidents of subsea pipelines in 2012 year amounted to $138,757 million, which gives an average damage per accident of about $1,11 million. According to [15], the total direct economic losses from accidents on US gas pipelines for the period 1986–2012 years amounted to $558,778 million. According to [16], the average damage from accidents at gas and oil pipelines is $\$10^4–\$10^7$, excluding the cost of gas losses. The actual gas losses reach 10^4 m^3.

As follows from the data presented, the frequency ranges of accidents for various water areas, pipelines and their operating conditions are within the range of $(10^{-6}–10^{-3})$ 1/km × year. Therefore, these values can be considered as the initial ones for substantiating the criteria for assessing the hazard of defects. Taking into account these frequencies and the amounts of damages presented above, the range of risks can be $\$10^1–\$10^4$ per accident. It should be noted that these values only include direct damages. Taking into account consequential damages, the risks can be significantly higher. It should also be emphasized that recently, risk assessments have taken into account not only the cost of restoring objects after accidents, but also the time of their restoration.

3. Brief of the problem of defects hazard assessing

The problem of assessing the safety of pipelines arose at the turn of the 50s - 60s due to the aging of pipeline systems in the United States. Later it became relevant for pipeline systems in other countries. The initial approaches to its solution were based on the methods of fracture mechanics, since the most large-scale accidents were caused by the development of cracks. For a number of reasons (the need for special tests, imperfection of models, the use of steels with increased crack resistance in pipes, etc.) they have not found wide practical application.

The pipeline transport development in the 1970s adduce three significant changes: pipeline systems swept the all world; the problem of ensuring the safety of pipeline systems, taking into account their aging, has become global; methods of in-line inspections (ILI) are become widely used. The ILI showed the presence of various types of defects in the pipes that reduce the efficiency of pipelines. Takin this in to account the defects hazard assessment began to occupy a special place in the security problem. To solve this problem, the methods ASME B31, APT1160, RSTRENG, DNV and others focused on the analysis of the most common defects in the form of corrosion damage [17] were developed. Parallel to this, the methods of breaking mechanics have developed and improved, which are reflected in the standards BS7910, API RP579, SINTAP.

Further research and development, sponsored by major international oil and gas companies (BP, DNV, Shell, Statoil, Total, and others), lead to the development of the Pipeline Defect Assessment Manual (PDAM). PDAM is based on a comprehensive critical review of available methods and full-scale pipe test results [18]. The scope of PDAM includes steel pipelines manufactured to API 5 L or equivalent national and international standards. The methods given in PDAM are applicable to defects in surface, underground and subsea pipelines. In these methods the following types of defects are considered: corrosion damage, scoring and marks, dents and corrugations, welding defects, delamination and cracking of the metal. These methods take into account the interactions of defects. The methods take into account the main and additional loads. At the same time, it should be noted that many PDAM methods are empirical, with a limited scope.

A significant drawback of PDAM methods is the use of a deterministic approach to defect hazard assessing. The dimensions of defects, loads and characteristics of the mechanical properties of steels are considered as deterministic, unambiguously given values. The partial safety factors used in the calculation methods are based on empirical data and are not directly related to the inevitable random variations of these parameters. Due to these circumstances PDAM methods are not combined with the developed concepts of Risk based performance management and Risk based Inspection (RBI).

Comparative analysis of methods for hazard assessment of pipeline defects allows us to draw the following conclusions:

1. For the bulk defects in the form of metal loss (corrosion) and dents the main parameters are the relative depth h/t and the relative length $l^2/(Dt)$. The defect size of around the circumference of the pipe is usually not taken into account. For the flat defects (crack, delamination) the main parameters are length and depth of the defect.

2. The calculated ratios used for the limiting sizes of defects differ in terms of the shape of approximation of the area A of defect cross-sections: rectangular ($A = hl$), parabolic ($A = 2h/3l$), combined ($A = 0.85hl$). It is not possible to single out a more accurate approximation on the available results of field tests of pipes with defects. Taking into account random variations in the shape and size of real defects, any approximation with an undefined error can be used.

3. Defect hazard assessments are carried out for given limit states function of pipes, defined as $\mathcal{L} = \Phi(P, Q, C_f, D, t, h, l)$, where Φ is a function of a given form, P is operation pressure; Q is external loads; C_f is the strength criterion of a pipe with a defect; D, t are pipe diameter and wall thickness; h, l are depth and length of defect [19].

In conclusion, it should be noted that pipeline defects are random, unique and complex in shape and their sizes are depend on the operating conditions and the properties of the external environment. The characteristics of defects cannot always be described by the current norms and calculation methods.

4. The concept assessing for hazard of defects by risk criteria

The above analysis shows that pipeline will invariably contain defects at some stage during its life. These defects will require a "fitness-for-purpose" assessment to determine whether or not to repair the pipeline. The full-scale tests of pipelines with defects and limit state functions method are used for such assessment. The limit state function method allows determining the limit size of defect upon reaching which the pipeline will fail. The limit state function $\mathcal{L}$ for pipe with defect can be write as:

$$\mathcal{L}\{P, Q, \sigma_f, D, t, l\} = l_r(P, Q, \sigma_f, D, t, l) - l_i = 0 \tag{1}$$

where P is operation pressure; Q is external loads; σ_f is fracture stress; D is outside diameter of pipe; t is wall thickness of pipe; l_r is allowable defect size; l_i is defect size in pipeline.

The defects sizes l_i are established during ILI. The allowable defects sizes l_r are determined by calculation methods by the specified criteria for the strength and durability of structures, taking in to account the operating conditions and the character of the mechanisms of deformation and destruction [6–10]. It should be emphasized that in these methods the sizes of defects l_i and l_r are assumed to be deterministic values.

In reality, the defects have inevitable random dispersion of sizes. For detected defects, these are caused by the random nature of the defects, as well as by statistical errors and the probabilistic nature of the operational characteristics (sensitivity and detectability) of non-destructive testing methods [16]. The dispersion of the calculate sizes of defects determined by statistical scattering loads, operating conditions and scattering of mechanical properties. A certain contribution to the possible dispersion of defect sizes is made by idealization of the shapes and schemes of defects. Taking this into account, instead of single-valued sizes in the calculations, it is necessary to use the probability densities distribution functions of defect sizes $f(l_i)$ and $f(l_r)$.

Using the functions $f(l_i)$ and $f(l_r)$ gives reason to believe that there are always nonzero probabilities P presence of defects with sizes l_i larger than l_r (**Figure 1**):

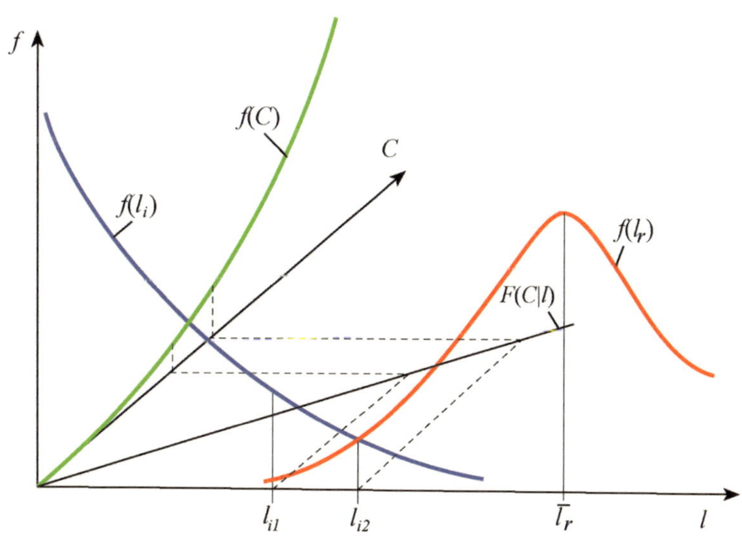

Figure 1.
Probabilistic scheme of the defects hazard analysis.

$$P(l_i > l_r) = \int_0^\infty \int_{l_r}^\infty f(l_i) f(l_r) dl_r dl_i \tag{2}$$

Exceeding the sizes l_r leads to some losses $C = F(l)$ due to the need to carry out repair operations or leakage and structural failure. Moreover, the larger the defect, the more significant losses can be. It should be emphasized that the losses are also random in magnitude, since it depends on the many technical and socio-economic factors.

Joint analysis of the probabilistic nature of defects, their hazard and possible losses leads to the concept of the admissibility of defects according to risk criteria [11, 18]. The essence of this concept is that the criterion condition for the admissibility of defects is represented in the form:

$$P(l_i > l_r) \times C(l) \leq [R], \tag{3}$$

where $[R]$ is the acceptable risk.

Assuming the defect size l_i as a fixed random variable from (3) we can obtain the following condition for the admissibility of a defect:

$$l_i \leq l_{[R]} \tag{4}$$

where $l_{[R]}$ is the size defect at which the risk R is acceptable.

Due to the unresolved problem of assessment and statistical analysis of losses, currently, sufficiently substantiated proposals for determining the allowable risk have not been developed. As a rule, losses are categorized into some qualitative classes: negligible, acceptable, unacceptable, etc. [19, 20]. Each class of losses is associated with a certain acceptable level of its probabilities $[R_f]$. Taking this into account, instead of (3), one can go to a simpler form of assessing the admissibility of defects by risk criteria, which does not require a direct assessment of damages, namely:

$$P(l_i > l_r) = \int_0^{l_i} f(l_r) dl_r \leq [R_f] \tag{5}$$

On this basis, similarly to (5), the following condition for the admissibility of defects can be written:

$$l_i \leq l_{[R_f]} \tag{6}$$

where $l_{[Rf]}$ is the size of the defect at which the probability of losses belongs to a given class.

Expressions (4) and (6), in fact, are a semi-probabilistic solution to problems (3) and (5), since they relate fixed random variables, one of which has a given probabilistic support.

5. Method for determining the allowable sizes of pipe defects by risk criteria

In this section the probabilistic methodology is use for develop a semi-probabilistic method for assessing the admissible sizes of defects in subsea inter-field pipelines based on risk criteria. The basis of this method are requirements of standards [7, 9]. The risk is defined as the probability R_f negative consequences of pipeline accident, the scale of which is determined by the hazard class. The proposed hazard classes (risk matrix) for inter-field subsea pipelines are presented in **Table 1**. Quantitative economic and environmental damage assessments are not considered here.

The suitability of the pipeline for operation is determined by three-level assessment of the allowable size of defects by risk criteria (**Figure 2**). The first, basic level, determines the allowable defect sizes by the strength characteristics of metal for pipelines exposed to the main loads - internal overpressure and hydrostatic external pressure. The second, extended level, determines the allowable defect sizes by the strength characteristics for metal, taking into account the effect on pipelines of additional longitudinal and bending loads. The third, special level, determines the allowable sizes of cracks, crack-like defects and delamination by the characteristics of crack resistance of the metal.

The calculations use information about: pipe sizes, location of the pipeline on the seabed, loads and impacts; the size, location and types of defects; mechanical properties, industry standard requirements, and pipe specifications.

The hazard of pipe defects depends on their shape and size. The sizes of defects are determined by their spatial coordinates $l = \{l_x, l_y, l_z\}$ (**Figure 3**). By shape the defects can be classified into volumetric and flat. For the volumetric defects the size $l_x \geq l_y \geq l_z$, for the flat defects the size $l_x \geq l_y >> l_z$. The defect hazard calculations usually use relative defect sizes $\tilde{l}_x = \frac{l_x}{\sqrt{Dt}}, \tilde{l}_z = \frac{l_z}{t}$. These relative dimensions are used in this technique taking into account the classification of defects shape.

The limit state function $\mathcal{L}$ for pipe volumetric defects may be write as:

for hoop stress

$$\mathcal{L}\left(P, D, t, \sigma_f, \tilde{l}_x, \tilde{l}_x\right) = \sigma_f \frac{2t}{D} RF\left(\tilde{l}_x, \tilde{l}_x\right) - P = 0 \tag{7}$$

for equivalent stress

$$\mathcal{L}\left(P, D, t, \sigma_f, \tilde{l}_x, \tilde{l}_x\right) = \sigma_f - \sigma_e RF\left(\tilde{l}_x, \tilde{l}_x\right) = 0 \tag{8}$$

Hazard classes	**Low**	**Middle**	**High**	**Very high**
Failure classes	Neglected	Uncritical	Critical	Catastrophic
Level of loss	Negligible environmen-tal and econo-mic impact. Pipeline repa-ir can be post-poned until the planned shutdown.	Short-term local disturbance of the state of the ecological environment and/or insignificant material losses. Unscheduled pipeline shut-down and repair.	Short-term damage to the environment and/or signify-cant economic damage. Unscheduled pipeline shut-down and repair.	Large-scale long-term environmental damage and large economic damage. Long shutdown and pipeline repair.
$[R_f]$	10^{-2}	10^{-3}	10^{-4}	10^{-5}

Table 1.
Hazard classes of fracture for subsea inter-field pipeline.

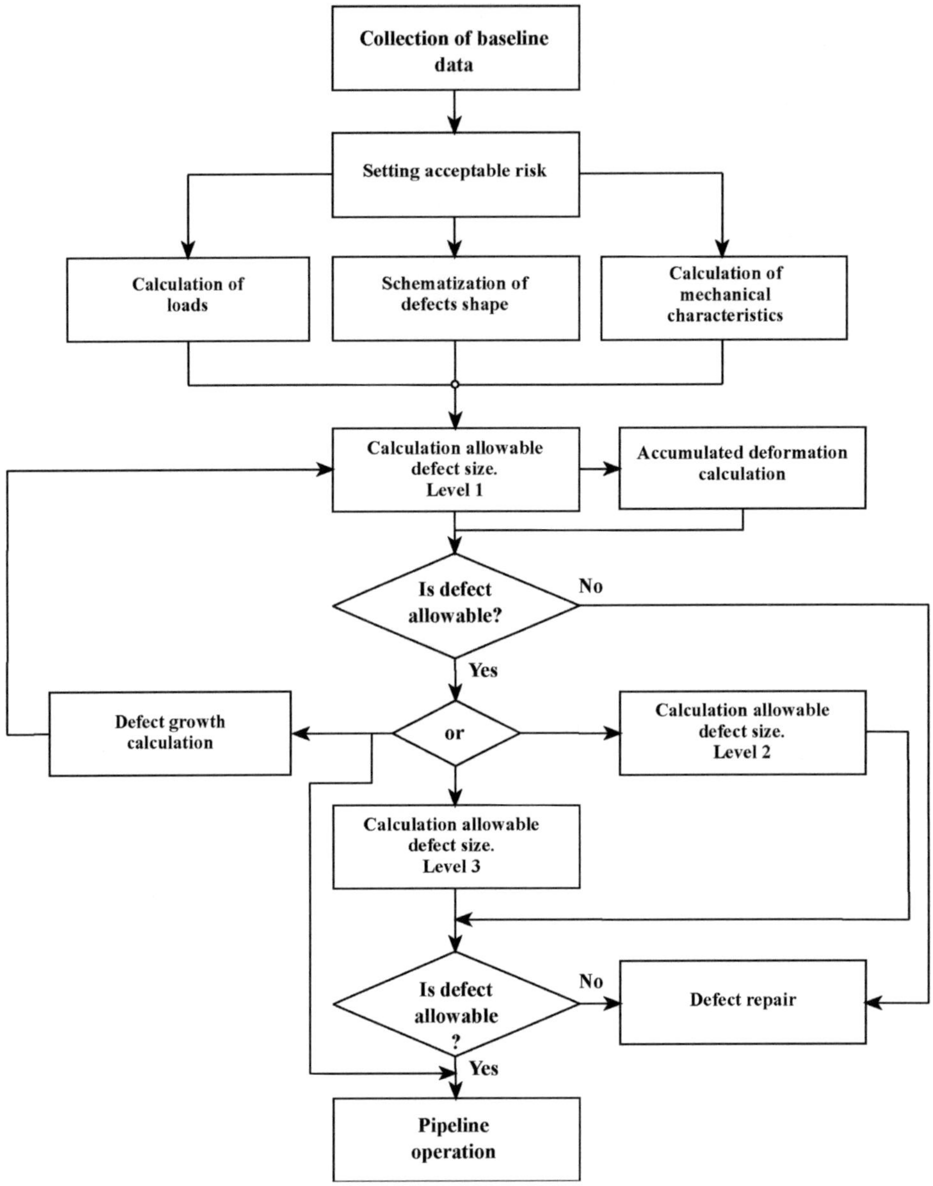

Figure 2.
Scheme for calculating the allowable size of defects.

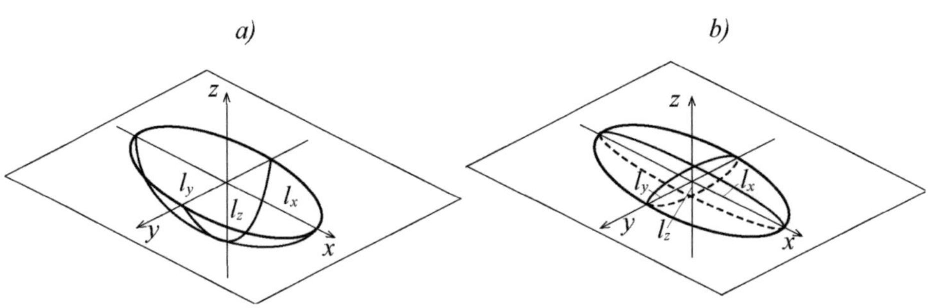

Figure 3.
Idealization of volumetric (a) and flat (b) defects shape.

where $\sigma_f = \min\left\{\frac{R_e}{\gamma_e}; \frac{R_m}{\gamma_m}\right\}$ is fracture stress; $\sigma_e = \sqrt{\sigma_h^2 + \sigma_l^2 - \sigma_h\sigma_l + 3\tau_{hl}^2}$ is equivalent stress; σ_h is hoop stress; σ_l is longitudinal stress; τ_{hl} is tangential shear stress; $RF\left(\tilde{l}_z, \tilde{l}_x\right) = \frac{1-\tilde{l}_z}{1-\tilde{l}_z/M\left(\tilde{l}_x\right)}$ is risk-factor of defect; $M\left(\tilde{l}_x\right)$ is Folies factor; γ_e, γ_m are partial safety factor.

If the components of limit state functions have a Gaussian distribution, then from the solutions of Eqs. (7) and (8) it is possible to determine the allowable sizes $\tilde{l}_z$ for given sizes $\tilde{l}_x$ with use partial safety factors:

for hoop stress

$$\tilde{l}_z \leq \frac{1}{\gamma_d} \frac{\sigma_f - 0.75\frac{\gamma_R PD}{t}}{1.1\sigma_f - \frac{\gamma_R PD}{2t}\frac{1}{M}} \tag{9}$$

for equivalent stress

$$\tilde{l}_z \leq \frac{1}{\gamma_d} \frac{\sigma_f\gamma_S - \sigma_e\gamma_\sigma}{1.1\sigma_f\gamma_S - \frac{\sigma_e\gamma_\sigma}{M}} \tag{10}$$

where γ_d, γ_R, γ_s, γ_σ are safety factors determined by the admissible of risk fracture $[R_f]$.

The safety factor γ_R is determined taking into account the admissible level of fracture probability $[R_f]$:

$$\gamma_R = \frac{1 - u_p\sqrt{V_f^2 + V_p^2 - \left(u_p V_f V_p\right)^2}}{1 - \left(u_p v_f\right)^2} \tag{11}$$

where u_p is the quantile corresponding to the probability $[R_f]$; V_f is the coefficient of variation of the fracture pressure; V_p is coefficient of variation of operation pressure.

The u_p quantile is set taking into account the accepted safety class of the pipeline according to **Table 2.**

The coefficients of variation of fracture pressure and operation pressures V_f and V_p are determined by statistical methods based on data for statistical scattering of the operation pressure, pipe metal mechanical characteristics, diameter D and wall thickness t of pipes.

The partial safety factor for the defect size γ_d is determined taking into account requirements [7] base on the value standard deviations $S_h/_t$ of the defect size (**Table 3**). The partial safety factors γ_s and γ_σ are set according to **Tables 4** and **5**.

The hazard of defect is determined by the design point position, given by the actual coordinates $\tilde{l}_z$ and $\tilde{l}_x$on the design diagram (**Figure 4**).

Hazard classes	Probability of fracture	u_p
I - Low	$\leq 10^{-2}$	2.33
II - Meddle	$\leq 10^{-3}$	3.1
III - High	$\leq 10^{-4}$	3.72
IV – Very high	$\leq 10^{-5}$	4.27

Table 2.
Values of quantiles u_p.

Hazard classes	Partial safety factor γ_d	
I - Low	$\gamma_d = 1.0 + 3.0S_{h/t}$	
II - Meddle	$\gamma_d = 1.0 + 4.0S_{h/t}$	$S_{h/t} < 0.04$
	$\gamma_d = 1.0 + 5.5S_{h/t} - 37.5S^2_{h/t}$	$0.04 \le S_{h/t} \le 0.08$
	$\gamma_d = 1.2$	$0.08 \le S_{h/t} \le 0.16$
III - High	$\gamma_d = 1.0 + 4.6S_{h/t} - 13.9S^2_{h/t}$	
IV – Very high	$\gamma_d = 1.0 + 4.3S_{h/t} - 4.1S^2_{h/t}$	

Table 3.
Values of safety factor γ_d.

Hazard classes	Low	Middle	High	Very high
γ_S	0.76	0.72	0.63	0.6

Table 4.
Values of safety factor γ_S.

Hazard classes	Low	Middle	High	Very high
γ_σ	1.12	1.4	1.5	1.6

Table 5.
Values of safety factor γ_σ.

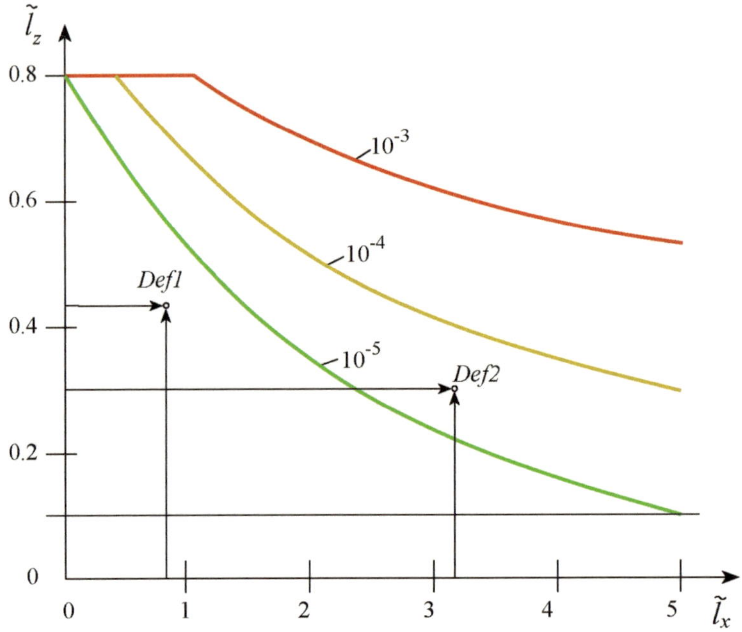

Figure 4.
Diagram for determining the allowable size of defects.

The assessment of the allowable sizes of flat defects (cracks and crack-like defects, delamination) in pipes is based on a Failure Assessment Diagram (FAD). The FAD concept combines the approaches of fracture mechanics to the analysis of brittle and quasi-brittle fractures, with the approaches of limiting analysis, which determines the conditions of ductile fracture of structural elements with crack-like defects. The fracture diagram is given by the following Equations [9, 10]:

$$f(L_r) = \begin{cases} \left(1 + 0.5L_r^2\right)^{-1/2} \times \left[0.3 + 0.7\exp\{-\mu L_r^6\right], & L_r < 1 \\ f(L_r = 1)L_r^{(n-1)/n} & 1 \le L_r \le L_r^{max} \end{cases} \tag{12}$$

Parameter L_r^{max} is calculated by the formula:

$$L_r^{max} = 0.5\left(1 + \frac{R_y}{R_m}\right) \tag{13}$$

Parameter μ is calculated as:

$$\mu = min\left\{\frac{0.001E}{R_e}; 0.6\right\} \tag{14}$$

Parameter n is calculated by the formula:

$$n = 0.3\left(1 - \frac{R_m}{R_y}\right) \tag{15}$$

The risk of fracture is taken into account by introducing safety factors for crack resistance and load:

$$K_r = \frac{f(L_r)}{\gamma_K}, L_r = \frac{L_r^{max}}{\gamma_L} \tag{16}$$

The values of safety factors γ_K and γ_L are taken according to **Tables 6** and 7.

The load parameter L_r is defined as the ratio of the working pressure P to the plastic flow pressure P_y of the section of a pipe with a crack, $L_r = P/P_y$. The plastic flow pressure P_y is determined taking into account the geometry and orientation of the crack in the pipe. The fracture toughness parameter K_r or J_r is defined as the ratio of the effective stress intensity factor K_{eff} or J-integral J_I to the fracture toughness characteristic of the material K_{mat} or J_{mat}:

$$K_r = K_{eff}/K_{mat}, J_r = J_I/J_{mat} \tag{17}$$

The effective stress intensity factor K_{eff} is determined taking into account the geometry and orientation of the crack in the pipe using fracture mechanics methods or by finite element method.

Hazard classes	Low	Middle	High	Very high
γ_K	1.41	1.73	2.23	3.16

Table 6.
Values of safety factor γ_K.

Hazard classes	Low	Middle	High	Very high
γ_L	1.5	1.8	2.25	3.0

Table 7.
Values of safety factor γ_L.

Based on results of the calculations a fracture diagram is constructed (**Figure 5**). The danger of defect is determined by the position of the design point, given by the coordinates (K_r, L_r) on the diagram. If the calculated point is inside the diagram, then the considered defect is admissible, with a given level of risk fracture.

The presented approach is applied in practice, taking into account the following provisions. The decision on the identified defects is made on the basis of all available information about their type, size and location, as well as the stability of the working loads and the operating conditions of the pipeline. Defects corresponding to the level of fracture probabilities R_f less than 10^{-5} according to the defect hazard diagrams are considered as allowable under the given operating conditions. Defects located in the zone of probability of destruction $10^{-5} < R_f \leq 10^{-4}$ are considered as potentially dangerous and are allowed for operation provided that there is a monitoring system and automatic limitation of internal pressure in the pipeline, and periodic non-destructive testing. Defects located in the zone of destruction probability $10^{-4} < R_f \leq 10^{-3}$ are considered dangerous and must be repaired in a planned manner. Defects located in the destruction probability zone $R_f > 10^{-3}$ according to the defect hazard diagrams are considered unacceptable and must be repaired immediately.

More promising is the transition from the described approach to probabilistic approach for determining allowable sizes of defects. Such approach is developed on the basis of taking into account probability density functions of distributions defects sizes $f(l)$. This approach assumes that the probability density $f(l)$ is a mixture of distributions of random variables included in the limiting state equation, and is approximated by the Weibull distribution [11]:

$$f(l) = \frac{\beta}{\theta}\left(\frac{l}{\theta}\right)^{\beta-1} exp\left\{-\left(\frac{l}{\theta}\right)^{\beta}\right\} \quad (18)$$

Substitution of expression (18) into (5) gives the following expression for the admissible size of the defect $\tilde{l}_z$:

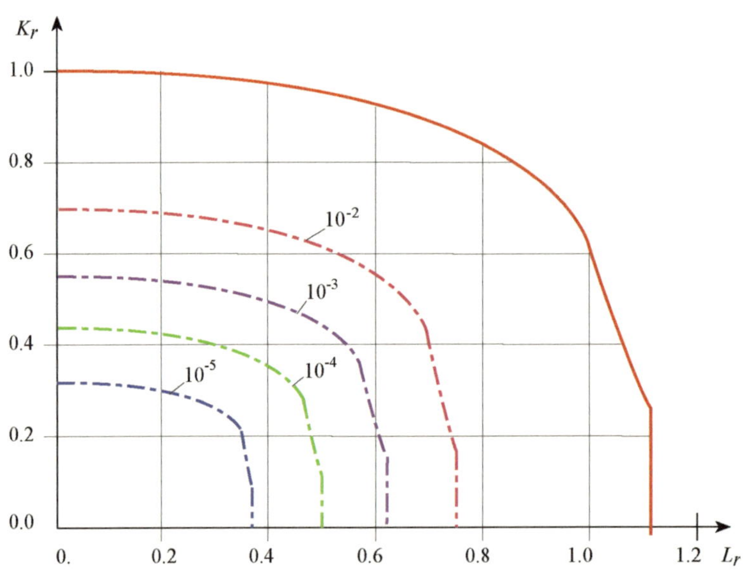

Figure 5.
Failure assessment diagram with risk level.

$$\tilde{l}_z \leq \theta \sqrt[\beta]{-\ln\left(1 - P_f\right)} \tag{19}$$

where P_f is the fracture probability corresponding to the given fracture risk R_f.

The parameters β and θ are related to the mean value μ_l and standard deviation S_l:

$$\mu_l = \theta \Gamma\left(1 + \frac{1}{\beta}\right), S_l = \theta \sqrt{\Gamma\left(1 + \frac{2}{\beta}\right) - \Gamma^2\left(1 + \frac{1}{\beta}\right)} \tag{20}$$

where $\Gamma(x)$ is Gamma function.

The mean value μ_l, standard deviation S_l, coefficient of variation V_l of the defect sizes can be determine based on experimental, calculated or literature data related to the subject pipeline.

The risk diagram can be constructed based on calculations for different probabilities P_f similar as shown above. The permissible defect sizes must be below the specified probability.

The presented probabilistic fracture model can be used to assess the risk of accidents based on the methodology of Probabilistic Risk Analysis (PRA). Features of solving this problem for subsea pipelines can be found in the works [21, 22].

6. Estimation of allowable defect sizes

As an example, **Figure 6** shows the results of a calculated assessment of the risk of metal loss defects in an inter-field subsea gas pipeline∅ 406.4 × 17.5 mm by risk criteria. The pipe material is steel X60 (R_y – 415 MPa, R_m – 520 MPa, E = 2.06 × 10^5 MPa, α_t = 1.1 × 10^{-5}). The operation pressure is 16 MPa. Temperature operation difference is ΔT = 50°C. The total number of detected defects is 916 pcs: h/t = from 20 to 39%, – 5 defects, h/t = from 10 to 19% – 82 defects, h/t < 9% – 829 defects. Of these, 16 defects are unacceptable according to the standard [2].

The presented results show, that three defects are located in a hazard area with a risk level higher than 10^{-3} and require immediate elimination. Two defects correspond to a risk level above 10^{-4} and can be corrected in a planned manner.

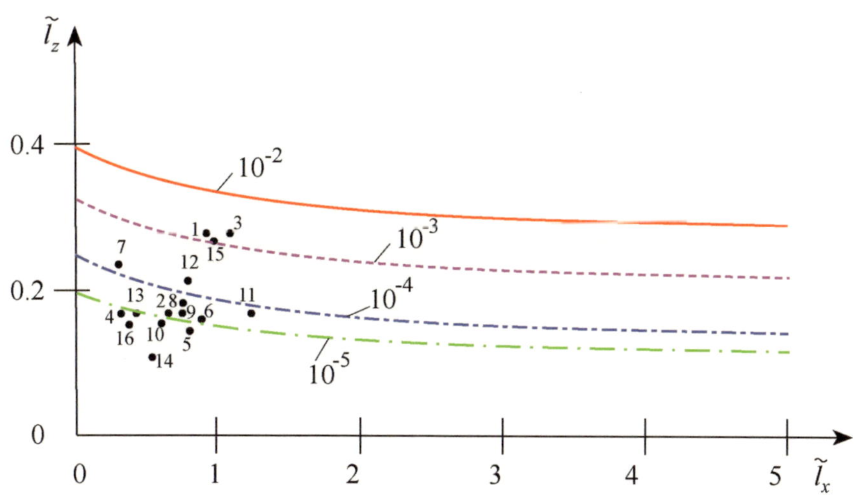

Figure 6.
The defects hazard diagram (with indicate defect numbers).

Five defects are in the risk zone 10^{-5}–10^{-4} and can be repaired as planned at a later date. Defects below the level 10^{-5} can be allowed for operation, provided that periodic non-destructive testing is carried out.

Thus, the proposed method provides a more flexible and more substantiated scheme for assessing the hazard of defects. On the one hand, this assessment takes into account the risk of accidents, thereby ensuring the required level of safety. On the other hand, it allows a more rational use of financial and material resources allocated for diagnostics and repair of subsea pipelines.

7. Conclusion

The paper discusses the possibilities of implementing the risk-based control method for inter-field subsea pipelines. The results obtained allow us to draw the following conclusions. Currently, there are a number of methods for assessing the hazard of pipeline defects based on deterministic approaches. Risk-based inspection provides greater opportunities for prioritizing, planning, justifying and evaluating the results of non-destructive testing. For the practical implementation of the risk-based control method, it is necessary to develop special probabilistic and semi-probabilistic calculation methods for assessing the hazard of pipeline defects taking into account random factors.

The proposed semi-probabilistic methodology is a development of the provisions of the DNVGL-ST-F101, SINTAP and DNV-RP-F116 standards. The novelty of the methodology lies in the justification of the safety factors through the level of failure probabilities corresponding to a given class of damage and loss. This opens up new possibilities for solving the problem of admissibility of defects in inter-field subsea pipelines from the standpoint of the concept of serviceability according to risk criteria.

Author details

Anatoly Lepikhin[1]*, Victor Leschenko[2] and Nikolay Makhutov[3]

1 Federal Research Center for Information and Computational Technologies, STC "Neftegazdiagnostika", Moscow, Russia

2 STC "Neftegazdiagnostika", Moscow, Russia

3 Mechanical Engineering Research Institute of the Russian Academy of Sciences, Moscow, Russia

*Address all correspondence to: aml@ict.nsc.ru

IntechOpen

© 2020 The Author(s). Licensee IntechOpen. This chapter is distributed under the terms of the Creative Commons Attribution License (http://creativecommons.org/licenses/by/3.0), which permits unrestricted use, distribution, and reproduction in any medium, provided the original work is properly cited.

References

[1] GOST R 54382–2011 Oil and gas industry. Subsea pipeline systems. General technical requirements. 274 p.

[2] ANSI/API 5L Pipes for pipelines. Technical conditions. 2007. 162 c.

[3] Volchenko VN. Probability and reliability of assessing the quality of metal products. Metallurgiya: Moscow; 1979 88 p

[4] Lepikhin AM. Reliability of norms technological defectiveness of welded joints. Preprint of the Computing Center of the Siberian Branch of the USSR Academy of Sciences №13. Krasnoyarsk. 1990:16

[5] Volchenko V.N., Konovalov N.N. Probabilistic calculations of norms of defectiveness of welded joints under high-cycle loading // Welding production. 1991. №8. P. 27–30.

[6] ND 2–020301-005. Rules for the classification and construction of subsea pipelines. 2017. 178 p.

[7] DNVGL-ST-F101 Submarine pipeline systems. Edition October 2017. 521 p.

[8] ASME B31G-1991 Manual for determining the remaining strength of corroded pipelines. 56 p.

[9] SINTAP (1999) Structural Integrity Assessment Procedure. Final Revision. EU-Project BE 95–1462. 231 p.

[10] BS7910:2013. Guide to methods for assessing the acceptability of flaws in metallic structures. 480 p.

[11] Lepikhin AM, Makhutov NA, Moskvichev VV, Chernyaev AP. Probabilistic risk analysis of technical systems structures. Novosibirsk: Nauka; 2003 174 p

[12] Carr P. A model to estimate the failure rates of offshore pipelines. Proceeding of IPC. 2010 https://www.researchgate.net/publication/267648527

[13] Christou M, Konstantinidou M. Safety of offshore oil and gas operations: lessons from post accidental analysis. Report EUR25646EN. 2012:60

[14] Offshore accident and failure frequency data sources-review and recommendations. RR1114. HSE Books, Derbyshire, 2017, 54 p.

[15] Girgin S, Kraussman E. Lesson learned from pipeline natech accidents and recommendations for natech scenario development. Report EUR 26913 EN. 2015:104

[16] Belvederesi C., Dann M. Statistical analysis of failure consequences for oil and gas pipelines // Int. Journal of safety and security Eng. Vol.7, №2 (2017), P. 103–112.

[17] Mustaffa Z., van Gelder P., Vrijling H. A Discussion of Deterministic vs. Probabilistic Method in Assessing Marine Pipeline Corrosions / Proceedings of the Nineteenth (2009) International Offshore and Polar Engineering Conference Osaka, Japan, June 21-26, 2009. p. 653–658.

[18] Cosman A., Hopkins P. An overview of the pipeline defect assessment manual (PDAM) / 4th International Pipeline Technology Conference 9-13 May 2004, Oostende, Belgium, p. 1–13. https://pdf4pro.com/amp/view/an-overview-of-the-pipeline-defect-1d123.html

[19] Mustaffa Z., van Gelder P. A Review and Probabilistic Analysis of Limit State Functions of Corroded Pipelines // Proceedings of the Twentieth (2010) International Offshore and Polar Engineering Conference Beijing, China, June 20–25, 2010. P. 626–632.

[20] Bay Y, Bay Q. Subsea pipeline integrity and risk management. New York: Elsevier; 2014. 405 p 978-0-12-394432-0

[21] Li X, Chen G. Zhu H. Quantitative risk analysis on leakage failure of submarine oil and gas pipelines using Bayesian network. Process Safety and Environmental Protection, Part A. 2016; **103**:167-173

[22] Liu C., Liao Y., Wang S., Li Y. Quantifying leakage and dispersion behaviors for sub-sea natural gas pipelines. Ocean Engineering. Volume 216, 15 November 2020, Article number 108107.

Section 3

Hazards and Impacts

Chapter 9

Analyzing the Cyber Risk in Critical Infrastructures

Marieke Klaver and Eric Luiijf

Abstract

Information and communication technology (ICT) plays an important role in critical infrastructures (CIs). Some ICT-based services are in itself critical for the functioning of society while other ICT elements are essential for the functioning of critical processes within CIs. Moreover, many critical processes within CIs are monitored and controlled by industrial control systems (ICS) also referred to as operational technology (OT). In line with the CI-concept, the concept of critical information infrastructure (CII) is introduced comprising both ICT and OT. It is shown that CIIs extend beyond the classical set of CIs. The risk to society due to inadvertent and deliberate CI/CII disruptions has increased due to the interrelation, complexity, and dependencies of CIs and CIIs. The cyber risk due to threats to and vulnerabilities of ICT and OT is outlined. Methods to analyze the cyber risk to CI and CII are discussed at both the organization, national, and the service chain levels. Cyber threats, threat actors, and the organizational, personnel, and technological cyber security challenges are outlined. An outlook is given to near future cyber security risk challenges, and therefore upcoming risk, stemming from (industrial) internet of things and other new cyber-embedded technologies.

Keywords: critical information infrastructure, cyber, risk, critical infrastructure, operational technology, industrial control systems, SCADA, internet of things, industrial internet of things, security, mitigation

1. Introduction

This chapter 'Analyzing the Cyber Risk in Critical Infrastructures' discusses the concepts of critical infrastructure (CI) and critical information infrastructure (CII), highlights the need for addressing the cyber risk to CI/CII, discusses methods and challenges in assessing the cybersecurity risk for CI/CII, and highlights upcoming cyber risk. This chapter brings together views on what comprises CII in the light of technological and societal developments, and how to analyze the cyber risk of CI and CII given the complexity of CI sector structures, dependencies, and service chains.

Following this introduction section, Section 2 introduces the concept of CII, its relation to the classical CI, and discusses the importance of analyzing the cyber risk to CI/CII. Section 3 discusses methods and challenges in analyzing the cyber risk to CI/CII both from the perspective of a single organization and across organizations e.g. across a CI sector or along a CI/CII service chain. Section 4 analyses the vulnerabilities and cyber risk of operational technology (OT) in CI. Section 5 discusses methods to analyze the cyber security risk across multiple organizations including

 IntechOpen

supply chains. Section 6 provides an outlook at new technological and regulatory developments and their possible impact on the cybersecurity risk for CI and CII. This chapter concludes with the conclusions in Section 7.

2. CI, CII, and the cyber risk

2.1 What is CI and how does that relate to CII?

The Council of the European Union has defined a CI as: "*an asset, system or part thereof located in Member States which is essential for the maintenance of vital societal functions, health, safety, security, economic or social well-being of people, and the disruption or destruction of which would have a significant impact in a Member State as a result of the failure to maintain those functions*" [1]. Currently, many states on the globe have defined a subset of their infrastructure services as CI using similar definitions for CI. Their aim is to guarantee the wellbeing of their population and economy by safeguarding the undisturbed functioning of the society under all hazards. A list of national definitions for CI can be found at [2].

To determine their set of national CI sectors, states use methodologies such as a national risk assessment (NRA) method [3, 4] or a risk-based approach in combination with a set of criteria [5]. CI are deemed critical at the national level if e.g. the number of casualties or the economic loss caused by disruptions exceed certain thresholds [6]. Most states recognize energy, telecommunications and internet, drinking water, food and health as CI sectors [7]. Within these CI sectors, states identified critical processes, products, and services at the *national level.* Depending on its economic structure, historic developments, cultural, and other factors, states may recognize other sectors as CI, e.g. social services, monuments and icons as shown by the webpage 'critical infrastructure sector' on [2].

In line with CI, CIIs comprise those ICT-based elements for which the disruption or destruction may – according to defined criticality criteria - have a serious impact on a state's society and its economy. CII is therefore defined by [8] as "*those interconnected information and communication infrastructures, the disruption or destruction of which would have serious impact on the health, safety, security, or economic well-being of citizens, or on the effective functioning of government or the economy*". Nevertheless, many states, which have defined their CI sectors, struggle in defining and accepting the concept of CII although the cyber risk to society extends beyond the classical set of CI sectors. Section 2.2 outlines the identification of CII and highlights why CIIs may extend beyond the currently identified national 'classical' sets of CI sectors.

2.2 Identifying CIIs

Alike the protection and resilience of CI, the protection and resilience of CII also starts with identifying CII. Many critical and essential services of our societies largely depend on the undisturbed functioning of underlying ICT and OT. According to [9], OT is "*the technology commonly found in cyber-physical systems that is used to manage physical processes and actuation through the direct sensing, monitoring and or control of physical devices*". The overarching term OT replaces many earlier notions for process control technologies to monitor and control cyber-physical processes (CPS): industrial control systems (ICS), distributed control systems (DCS), energy management systems (EMS), supervisory control and data acquisition (SCADA) systems, industrial automation and control systems (IACS), and process automation (PA) [10]. To mention a few applications of OT: the generation, transport and distribution of various modes of energy, refinery processes, building automation systems (air-conditioning, elevators, fire alarm system), physical security access (locks, gates, cameras), laboratory analysis systems, tunnel safety systems, harbor cranes, and automatic guided vehicles (AGV).

Identifying the ICT- and OT-based services that are critical for a state proves to be complex. Most states struggle in clearly understanding and defining the information infrastructure components of critical processes to the state and its population. CII elements and services are notoriously more difficult and complex to demarcate and define than CI, both technically, organizationally, and from a governance point of view.

CII elements tend to be more interwoven and tend to hide within a CI, in cyber-physical processes, and in stacks of information-based services. The speed of innovation and uptake of new digital technologies in processes that evolve into critical processes to the society is high. Obviously this is complex as the critical ICT- and OT-based functions and services hide themselves (1) in the IT-sector (telecommunication and internet), (2) classical sector-specific CIs (**Figure 1**), and (3) even beyond these established domains.

According to [11], CII comprise:

1. Critical elements and services of the ICT sector, for example mobile telecommunication data services, internet exchange points, domain name services, certificate infrastructures, and Global Navigation Satellite Systems such as Galileo, BeiDou, and GPS for Position, Navigation and Timing (PNT) services.

2. Critical information, communication, and operational infrastructure elements- ICT and OT- in each of the CI. This may include e.g. critical financial transaction systems in the financial sector, critical logistic information systems, and OT which monitor and control critical cyber-physical systems such as in gas transport, harbors, railways, healthcare, and refineries.

3. The products and services of manufacturers, vendors and system integrators which are used across multiple CI sectors, nationally and internationally, whose vulnerability or common cause failure may negatively impact the proper functioning of CII and the CI that they are a critical element of.

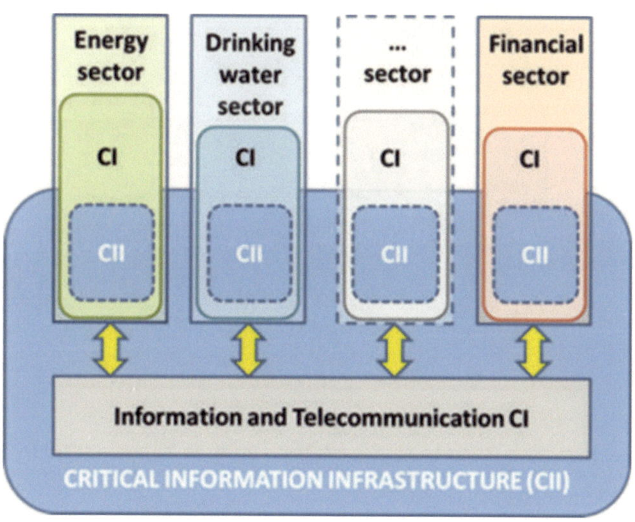

Figure 1.
Critical information infrastructure (source: [11]).

4. Critical ICT- and OT-elements and services beyond the established CI domains mentioned under (1) to (3) above. Such elements are often operated by organizations outside the classical ministerial supervision and/or regulation, may be physically located outside a state and or operated by foreign operators.

The extent of the nationally identified CII largely depends on the maturity and critical use of digital technologies by and in states (**Figure 2**). As a basis, essential CII elements include the ICT-based elements of the classical CI services such as electricity generation or drinking water. Digitally more advanced states have defined CIIs which have major elements outside the classical set of CIs. Due to the international nature of CII, the governance of CII protection and resilience extends beyond national borders and relies on international collaboration. Due to the increased role of ICT and OT in almost all other CI (e.g. cloud services, smart cities, smart grids), defining the CII requires cyclic updates to capture the dynamics inherently linked to ICT- and OT-based systems and networks. This process is complex due to the dynamics of the dependencies, and also to the sometimes-hidden nature of these dependencies, think e.g. on the dependency of electricity networks on the availability of precise timing and communication networks [12].

The EU, for instance, recognizes the need to secure both CI and CII in its European directive on security of network and information systems (NIS) [13]. The directive requires a higher level of cyber security by the operators of specific CI services in the energy (electricity, oil, and gas), transport (air, rail, over water, and road), banking, financial markets, health, drinking water supply and distribution, and digital infrastructure sectors. The non-classical CI 'digital infrastructure' comprises internet exchange points (IXs), domain name service providers (DNS), and top-level domain (TLD) name registries. EU Member States require by law that other national CI operators adhere to the same security requirements as well. Moreover, the NIS directive recognizes another set of CII operators: the digital service providers (DSPs). DSPs operate online marketplaces, online search engines, and cloud computing services when their operations exceed a certain size.

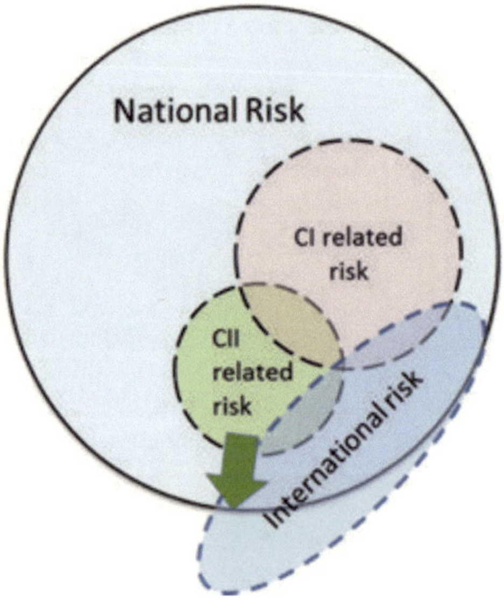

Figure 2.
Critical information infrastructure protection (CIIP): All activities aimed at ensuring the functionality, continuity, and integrity of CII to deter, mitigate and neutralize a threat, risk or vulnerability or minimize the impact of an incident. (source: [11]).

Moreover, the EU implicitly recognizes electronic identification and trust services for electronic transactions as CII in [14]. However, it should be noted that most EU states do not recognize their key registers on population, land, addresses and buildings, commercial companies, topology, and vehicles as CII [7].

The USA recognizes as life critical embedded systems as CII beyond the classical CI sectors: medical devices, internet-connected cars, and OT [15]. Other states, alike Australia, are in the process of identifying their CII.

The high dynamics of technological developments and subsequent societal use of ICT- and OT-based services, makes the identification of CII complex. What seems to be a new toy may become embedded in critical societal processes shortly. On the other hand, earlier critical services such as text messaging phase out while being replaced by newer mechanisms such as Whatsapp. Risk analysis and mitigation may be complex given (1) the ICT- and OT-technological dynamics, (2) the continuous shifts in the threat spectrum, and (3) new CII services often operated by new, non-traditional operators (e.g. cloud services) which do not fit automatically in the governance structures of states.

2.3 Why considering the cyber related risk to CI and CII?

The most feared phenomenon by states is the cascading effect due to dependencies between CIs and CIIs. When one CI or CII is disrupted or destroyed, cascading disruption(s) may occur through the dependency of other (critical) infrastructure(s). Another important risk factor to CI and CII is a common cause failure: "*a failure where the function of multiple infrastructures is disrupted or destroyed by the same cause or hazard affecting these infrastructures at the same location or area in the same time frame*" [2]. Common cause failures may for instance be triggered by extreme weather, flooding, wildfires, and common use of the same vulnerable ICT or OT application, software, or equipment.

In modern societies, the (cyber) risk to society and the economy due to inadvertent and deliberate CI/CII disruptions and cascading and common cause phenomena increases due to:

- The diminishing governmental control over classical CIs and CIIs due to liberalization and privatization of their operations.

- A more economic-based risk approach by CI and CII operators aiming for improved efficiency, productivity, and organization performance, as compared to a more societal risk-based approach by the earlier public CI/CII operators.

- The fast appearance of new ICT-based services that are perceived essential or even critical by society even before government considers them as being CII.

- The perceived critical use by citizens of new stacked services which make the underlying ICT-infrastructure critical, e.g. the mobile e-payment infrastructure.

- Urbanization which stresses the, often aging, CIs to the limits of their design capability and capacity.

- The increased dependence of CI on ICT and the hidden nature on some dependencies, see for instance [12] for possible cascading effects of disruptions of time synchronization services in electrical power networks.

- The increased use of vulnerable ICT and OT for the monitoring and control of CI operations.
- Complex dependencies of CI/CII services and the risk of cascading failures.
- The increased dependence of industries and the population on undisturbed CI and CII services. They expect and require a high level of CI/CII resilience, basically an undisturbed service 24 hours per day, all year around. Modern societies and its population cannot cope anymore with CI/CII service disruptions that affect a large area and have a long duration, citizens and businesses have no plan 'B'.
- The increased level of cyber-attacks by state actors [16] and other types of actors [17] deliberately performing (cyber) attacks on CIs and CIIs in support of their political and financial objectives. See e.g. the warning in [18].
- Vulnerabilities in commonly used ICT- or OT-applications and systems being the source of a common cause failure, e.g. a common vulnerability in a popular application may lead to vulnerabilities in many organizations simultaneously, see e.g. the Dutch national cyber security centre (NCSC) warning for a Citrix vulnerability [19].
- The high dynamics in vulnerabilities of ICT- and OT-applications and -systems.

Therefore, the analysis and mitigation of the cyber risk in CIs and CIIs pose major challenges to states and their operators of essential services.

3. Assess the cyber security risk in CI

3.1 CI, CII and risk analysis

Risk analysis is defined by the EU as the "*consideration of relevant threat scenarios, in order to assess the vulnerability and the potential impact of disruption or destruction of critical infrastructure*". [1] The Council of Europe's European Centre of Technological Safety (TESEC) defines risk analysis as: "*the determination of the likelihood of an event (probability) and the consequences of its occurrence (impact) for the purpose of comparing possible risks and making risk management decisions*" [20]. Identifying the cyber threat scenarios and vulnerabilities related to CIs and CIIs is an important element of the sectoral, national, and wider CI and CII protection and resilience policies and frameworks [13, 5–7]. Managing the characteristics requires thorough and regular assessments of the cyber risk for CIs and CIIs, both at the level of a single CI/CII operator, across a CI/CII sector, across CI/CII chains of services, and at the national level.

Risk assessment (RA) is "*the combination of vulnerability analysis and risk analysis*" leading to the "*determination and presentation (usually in quantitative form) of the potential hazards, and the likelihood and the extent of harm that may result from these hazards*" [20].

Risk analysis, vulnerability analysis, and, subsequently, RA are therefore important elements of the CI/CII protection and resilience efforts. Moreover, the risk management (RM) process for CI and CII should not only cover the business perspective of the risk but should also cover the societal impact of the risk: what risk does society faces when a large-scale disruption occurs? This requires RAs at multiple levels of aggregation, each with a different objective:

- An operator of essential services (CI or CII) will primarily use RA to obtain an overview of possible risk factors that can harm its business objectives and profits. Legal requirements will be a mere boundary condition to this process. The cyber risk is just one aspect which is balanced with other risk aspects such as e.g. technical failure, lack of key personnel due to a pandemic, and adverse regulation.

- A RA at the CI/CII sector level will primarily focus on the resilience and reputation of the whole sector considering the individual mitigation measures taken by the operators within the sector. E.g. what is the risk of diminished trust by the population in e-banking?

- A RA for a specific CI or CII service which depends on a chain of intermediate services supplied by multiple service operators. The operator of the (end) service will primarily focus on the resilience of the whole service chain and the disruption risk due to failing or disruption of one or more of the intermediate services. The analysis will consider the individual resilience measures taken by the individual operators and the residual risk for the service chain.

- A RA at the national or regional level will primarily focus on risk with societal impact and will take a wider range than just CI and CII. A national or regional RA will e.g. also consider the risk of a pandemic outbreak or a large-scale flooding and will balance the outcomes with the cyber risk to CIs and CIIs. To assess this risk, various states use a National Risk Assessment (NRA) method to establish a balanced national risk view including the cyber risk, see e.g. [3, 4, 21–23].

Due to the importance of CIs and CIIs for societies, CI and CII sectors increasingly must analyze and assess their (cyber) risk regularly and systematically based on sector-specific regulations either imposed by the national regulator, e.g. [24], or through sector initiatives, e.g. the Basel III regulatory framework for the bank sector. The implementation of the EU NIS directive as discussed above requires CI and some of the CII operators to regularly perform RAs as a basis for their cyber security measures. RM is also a key element in the NIST framework [25].

Moreover, these CI and CII operators should be prepared to perform a quick reassessment of the cyber risk, mitigations, and the residual cyber-related risk in case a new cyber vulnerability or cyber threat comes to the fore.

3.2 Assessment of cyber risk by a single CI operator

The basis for the protection of CI lies in a strong RA at the operator level. For RA at the company level, including CI and CII operators, many methods and standards exist. Most of these methods are in line with the ISO 31000 series of RM standards [26]. For the IT-environment, ISO/IEC 27005 [27] provides the RM and risk mitigation background as part of the ISO/IEC 27000 series that assist organizations to implement information security management based on a set of terms and definitions [28] and security controls [29, 30]. For the OT-environment, security control frameworks with similar security control sets exist, e.g. [31, 32]. Although these security control frameworks are often sector specific, they can be mapped on common structures or frameworks, see e.g. ENISA and NIST [25, 33].

One of the important factors to cover in a RA of CI/CII is the risk of ICT/OT as a vulnerability that may cause disruptions of CI/CII. This may involve the risk of technical failure or human mistakes, but also the cyber risk of malicious attacks. Given the criticality for states, even hybrid conflicts affecting CIs and CIIs are envisioned, see e.g. [34, 35]. An early example is the Crimea conflict. On December 23, 2015, Ukrainian power companies experienced unscheduled power outages impacting many customers in Ukraine. In addition, there have also been reports of malware found in Ukrainian companies in a variety of their CI sectors [36].

Section 4 below specifically focusses on the cyber risk factors related to OT.

3.3 Assessment of the cyber risk across organizations

A RA for a specific CI sector is feasible, as was shown by the EUropean Risk Assessment and COntingency planning Methodologies for interconnected energy networks (EURACOM) project [37]. This approach extended the EUropean Risk Assessment Methodology (EURAM) [38] with contingency planning. In particular, chapter 4 of the EURACOM report discusses the cyber threats to the energy CI sector. The methodology is based on a common and holistic approach (end-to-end energy supply chain) for RA, RM and contingency planning across the power, gas, and oil CI subsectors.

The seven steps of the EURAM RA methodology are shown in **Figure 3**. The methodology scales from the department level to the operator level, to the CI or CII sector, and national level. Moreover, the methodology may embed the results of other RA methodologies. Risk which cannot be dealt with at a certain level may be input to the next higher level of abstraction. For example, the risk implications of a pandemic or a state actor cyber-attack to a nation cannot be managed alone by a CI operator and must be off-loaded to and managed at the national or even supranational level.

3.4 Challenges to assess ICT/OT risk across organizations

Although methods and approaches exist to perform RA across organizations. (e.g. a CI/CII sector or a service chain) some practical challenges exist:

- *The risk attached to ICT and OT elements across CI/CII-chains.* Certain CI/CII services are composed of a set of (chained) ICT and OT elements provided and operated by multiple operators. The criticality of certain elements to a CI or CII may be unknown to its operator; therefore, its protection has less priority than required from the national CI protection (CIP) or NIS point of view. It is a challenge to identify such critical elements and to assess the risk attached across the chain. In support of this type of assessments, new methods have been proposed, e.g. the RA method suggested by the Dutch cyber security council which requires the collaboration of all organizations in a supply chain to collectively assess the risk and define the appropriate security controls [39].

- *Identifying the risk related to critical elements in various CI/CII*: Some ICT and OT products are widely used across many CI and CII sectors and other organizations. The cyber risk attached to a systemic failure or vulnerability of such

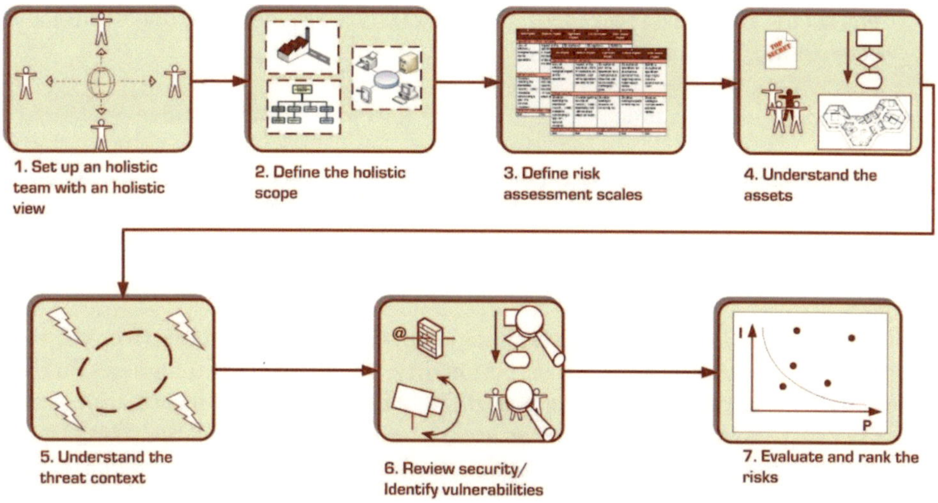

Figure 3.
The EUropean Risk Assessment Methodology (EURAM) approach (source: [38]).

a product may be large, e.g. a vulnerability in Microsoft Windows systems or in commonly used OT systems. Such a vulnerability may lead to a high level of risk at the national or even the international level. This risk is difficult to assess since it requires a detailed and well-maintained asset inventory of systems and applications used by each CI/CII operator.

- *The international nature of part of the CII:* Assessing the risk and taking mitigating measures for CIIP might be troublesome when the CII ownership, operations and or (operational) jurisdiction are beyond one's national border. Conflict of interests, legal requirements, and procedures may occur. For example, a cloud server operator having its operations in state B should report a cyber security breach to the national authority in that state. However, state A may have made regulation that each CII operator should report security breaches within 24 hours to them. When a CI operator in state A uses such a cloud service, the cloud service could have been designated as CII thereby imposing regulation on the cloud operator in state B. Such cross-border CII issues arise with the diverse national implementations of the EU NIS directive [40], and other CII-related laws. The new EU security strategy intends to address these issues [35].

These challenges lead to the necessity to perform RM not only at the company level but also across the service chain, and at the sector and national levels.

4. Assessing the OT risk

4.1 OT threats and vulnerabilities

To identify the main threats and vulnerabilities for the OT environment, a structured approach will be used in distinguishing multiple layers. Threats to OT may occur at multiple layers as defined by [41]:

- The governance layer.

- The socio-technical layer comprising the OT/ICT architecture, the technology, networking, and human factors.
- The operational-technical layer including (3rd party) maintenance.

According to [42], a threat to OT is the "*potential cause of an unwanted incident through the use of one of more OT, which may result in harm to individuals, a system, an organization, critical infrastructure and vital societal services, the environment or the society at large*".

The governance layer. At the governance layer, the first threat stems from the fact that OT is technically embedded in functionality. The management focusses on the functionality, e.g. provide drinking water. Therefore, many chief information security officers (CISOs) or equivalent executive level responsibilities largely neglect the cyber risk to OT which at the same time is a major risk to the functioning of the whole CI.

Moreover, there is major cultural difference between the IT department and other departments which use OT as part of the 24/7 functionality of their CI services. In addition, the IT department often has the cyber security mandate for the whole organization. "IT" develops the organization-wide cyber security policies (e.g. authentication and password policy, patch and anti-malware policies). Protection of the integrity, confidentiality, and privacy of information is a high priority. Therefore, "IT" may disrupt its operational services when required to install urgent patches. In their mindset, "IT" is key to the business of the whole organization; "*OT is just the department of grease, pumps, and valves, isn't it?*"

The OT department on the other hand optimizes the control of the physical processes and are less concerned with cyber security. Most often, "OT" has to use of the networks managed by "IT" for wide area connectivity and remote access. "IT" even may state the company-wide cyber security policy to comply with specific cyber security management standards such as the ISO/IEC 27000-series [28]. "OT" has to adhere to those policies while such cyber security standards and good practices have not been developed for a 24/7 operational environment. For example, blocking an account after three subsequent login errors is of no help when an operator needs to change production settings in the middle of the night during an operational crisis. Such dissimilar needs, policies, and service expectations between "IT" and "OT" can be a source of conflicts. Governance of OT security therefore requires efforts by all involved to bridge the gap between the ICT and OT domains.

Another governance level threat is that the economic depreciation of OT is often equal to that of the OT-controlled system, e.g. a water purification unit. Therefore, very aged control system components such as a 486 Windows/XP system still operate hidden in cabinets. They still control metros, sewage systems, and so on.

In other situations, the renewal of OT will be a long-term process where the upgrade will be performed (sub)process by (sub)process. This means that the central system control must cooperate with both new and legacy OT. Mixed configurations mean that cyber security measures cannot be activated at all or can only be effective on and between the new OT-systems and applications.

"*No worry about cyber security of OT, the processes still can be controlled manually*". At least management holds that view neglecting that the same management considerably reduced the experienced workforce able to manually operate the CI system. Therefore, an OT-disruption for longer than a couple of hours inevitably brings down the OT-controlled CI/CII services to society.

The socio-technical layer. At the socio-technical layer, [42] identifies a number of threats to the undisturbed functioning of OT-controlled CI processes, and therefore to the continuity, integrity and safety of physical processes. For example:

- Lack of cyber-security awareness of operators and other people operating and maintaining OT-controlled processes. No specific cyber-security education and training is part of their curricula.

- In the process control environment, it is not unusual that employees have been employed for many years. The risk of sabotage activities by disgruntled and dismissed employees is large. Many cases can be found in the media, e.g. the Maroochy water breach, and a sabotaged leak detection system of the Pacific Oil platforms and pipelines near Huntington Beach, USA. A risk which is not new: insider OT sabotage occurred already in the 90's, see e.g. [43].

The operational-technical layer. At the operational-technical layer, [42] identifies OT-specific threats including:

- The SCADA (and similar) protocols were designed in the 60's with a no threat, benign, closed operating environment in mind. Such protocols are not robust against any serious cyberattack. Applying such protocols now on top of TCP/IP increases the risk even more. A malformed packet may crash or lead to a dementia paralytica of process logic controllers as was shown by [44].

- The use of old technology and legacy OT, for reasons mentioned above, requires the need for personnel still knowing all ins and outs of twenty year or older OT as well as current technology. The old OT has no security-by-design. Moreover, old OT has too limited CPU and memory resources to run a malware protection package or encryption; the addition may break the critical process monitoring and control cycle. Moreover, a new plug-compatible board to replace a defective one may introduce new vulnerable functionality that is attractive to cyber attackers.

- In standard "IT" communications, temporary blocking of transmissions is accepted. In the OT-environment, however, not timely received status information from a process or a delayed control command may cause irreversible effects in the physical environment.

- OT systems may directly or indirectly be connected via remote operations or maintenance with the internet. Shodan [45] and similar search engine tools show ample OT-equipment that are directly accessible via the internet.

- System maintenance of OT in CI requires a lot of efforts due to the sheer size of the number of components. Password management policies, e.g. replacing passwords regularly, conflicts with the 24/7 operational continuity. CI sectors have agreed to good practices for patching and anti-malware signature updates but struggle with applying them, e.g. to apply security critical patches within a week after publication; all other patches to be applied during the next scheduled maintenance slot [46, 47]. In practice, patches are applied some three-quarter years after they became available and anti-malware signature files are updated after weeks if not months. "*If the controlled process works, do not break it*" is used as an excuse. Therefore, the risk of unauthorized exploitation of OT in CI sectors is high.

- Third party maintenance engineers are often given unrestricted and unmonitored access to key processes 24/7. Incidents have shown that third party employees cannot always be trusted.

4.2 Assessing the assurance of equipment and applications

A complex element in identifying the cyber risk in CII operations is assessing the risk in the wide variety of hardware and software CI operators use. Most CI/CII operators use ICT and OT from a multitude of suppliers, partly being global players. The hardware and software may contain hidden vulnerabilities. A CI/CII operator should try to ensure a high level of security of their own hardware, software, and services, and of those that are procured from suppliers. Organizations should adopt a security lifecycle approach to enhance the safe and secure functioning of their ICT elements. The security lifecycle comprises the acquisition, installation, system integration, operations, maintenance, upgrading, and decommissioning phases. When CI/CII operators are dependent on ICT and OT suppliers, system integrators, and third-party maintenance companies, they should have contractual agreements and measures in place to ensure that the resilience is up to par with the security requirements of the CI/CII organization. Based on the efforts of each organization, the use of cyber security standards and frameworks may increase the level of resilience across the chain. Examples of this approach are the third-party security requirements included in cyber security standards and frameworks [25, 29, 30, 32].

Assessing the level of assurance of each ICT/OT element, proves to be a challenge for an individual organization. Therefore, many organizations require support from their government, e.g. in certification of certain equipment. Recently, the EU Cyber Security Act [48] provides a framework structure for certifications, which is being taken up by ENISA and several of the European states although a number of challenges is perceived [49, 50].

4.3 Assessing the risk for the OT environment

The above-mentioned characteristics of OT systems, makes it necessary to include the following steps as part of the RA process:

- Use a multi-disciplinary team to assess the holistic risk to cyber. The team shall include those involved with general IT security, OT security, physical security, electronic security, security of services and supplies by utilities and third parties (e.g. power, external telecommunications, cooling), human resources (e.g. personnel security and safety).

- Collaborate with government organizations and relevant computer incident response teams (CSIRTs) on threat information and on assessing the risk to OT-equipment, software, and (tele)communication means.

- Identify the ICT and OT systems and networks that are critical to the key operational processes of the CI operator.

- Assess the impact of a disruption of ICT and OT to the CI service(s).

- Identify the connections with outside networks.

- Identify the external dependencies including third parties.

- Identify legacy systems that may pose additional vulnerabilities.

5. Assessing cyber security risk across CI/CII chains

Section 3.4 discussed the challenges for risk analysis across organizations in CI/CII chains. There exist several methods that support risk analysis across a chain of organizations which provide critical or essential services. There are, however, many challenges in applying such methods as is shown in Section 5.2.

5.1 Methods to assess the cyber risk across chains

Due to the specific characteristics, there is a need to perform RM not only at the company level but also perform a collaborative assessment across CI/CII service chains. There have been some studies that aim to establish a method for assessing the cyber-security risk across chains of CI/CII operations [38, 39].

The Dutch chain analysis method [39] has been developed by a set of CI operators in the energy sector. It was their believe that organizations in a supply chain together are in the best position to define and deploy appropriate controls and initiatives to reduce any cyber security risk themselves. The method aims to provide insight into the cyber security risk within a supply chain. It uses a layered approach to provide insight into the risk that arise from the ICT/OT systems and their interconnections as well as the potential risk that may pose to the chain of business processes of organizations. The identified risk in the business processes can ultimately disrupt the continuity of the entire supply chain of one or more critical or essential CI/CII services. By combining and merging the identified risk in business processes per organization, which should include their own third-party risk to these processes, the overall risk to the supply chain can be assessed (see **Figure 4**).

The aforementioned EURAM/EURACOM method uses a similar approach by combining three components to assess risk at an aggregated level, based on RAs by the individual organizations and is based on embedding lower level RA results by mapping the identified risk at the higher level [38].

Note that due to the hidden nature of ICT and OT within CI and CII, RM across the chain requires a large effort and a combination of expertise by all stakeholders

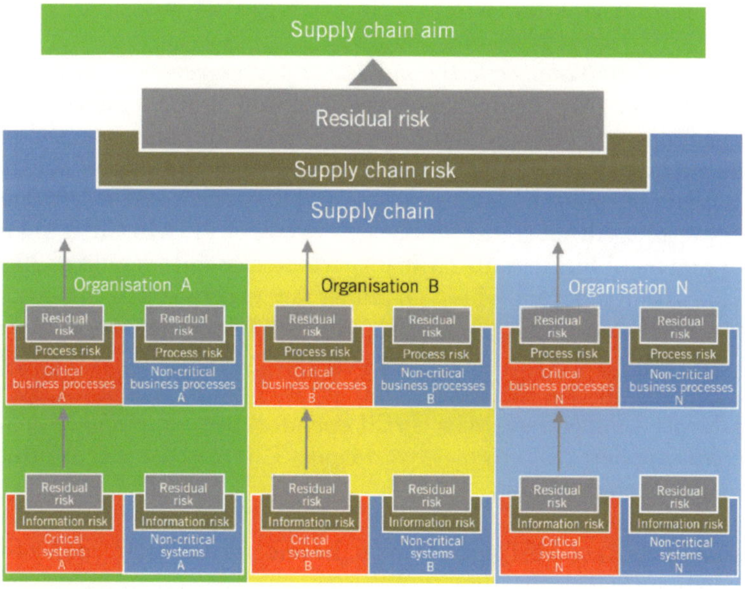

Figure 4.
Visualization of the Dutch supply chain risk management method (from [39]).

to assess this risk and define appropriate mitigating measures as is highlighted by the aforementioned Dutch supply chain RA pilot [39]: "*Providing insight into the cyber security risk within a supply chain requires a level of commitment of all organizations involved. It is paramount that in addition to the availability of adequate resources sufficient trust exists between organizations to share sensitive information among each other*."

5.2 Challenges to assess the cyber risk across CI/CII chains

In safeguarding CI and CII, cyber risk mitigation plays an important role. Cyber risk mitigation approaches comprise legal frameworks [13], the implementation of mostly non-CI/CII specific cyber security frameworks for ICT and OT [25, 29–32, 51], the sharing of cyber security information [52, 53], and a collaborative approach. The incentive for collaborative action to the cyber risk at the sector level and across service chains is clear. Resources are scarce and can be optimized by collaborating. Due to the interconnectedness of CI and CII, all organizations in a sector or service chain suffer when one weak link exists and fails, making a joint approach a necessity. Although many initiatives exist, the uptake of these initiatives is sometimes less than planned. Although there are methods available to assess the cyber risk across a CI chain, there exist challenges to apply those methods. Some of the factors that may prove a barrier in the adaptation of these methodologies are:

- *Different RA methodologies used by individual organizations:* Collaboration of RA across chains requires information sharing and discussions on the results of RA for the individual organizations. The sharing of information on the RA may be hampered when different methodologies are used. Although there are ways to overcome this, see e.g. [38], this requires some additional effort by the participating organizations.

- *Scarce resources:* Cyber security is a domain where expertise is still a scarce resource. When large scale incidents occur that would benefit from cross-organizational collaboration, many of the personnel needed will be taken-up by high-priority activities within their own organizations.

- *Difficulties in establishing effective public and private partnerships:* collaboration across the chain may require a close collaboration between public and private organizations, e.g. on information sharing on threats and vulnerabilities. While public-private partnerships (PPPs) are a popular form of collaboration in a number of states, in practice we see that they often lead to less than satisfactory results. Although the precise failure rate of PPPs in CIP is unknown, in the context of business-to-business partnerships failure rates of 30% up to 80% have been reported. This high failure rate may be based on tensions inherent to a PPP. Some balancing mechanisms are needed to overcome the inherent tensions [54].

- *Cross-border collaboration:* Most CI/CII operators use equipment of many different suppliers that originate worldwide. This may hamper information sharing and collaboration.

- *Legal barriers:* Anti-trust legislation on the one hand, and Freedom of Information (FOI) legislation on the other hand may create barriers to collaborate and exchange information between organizations [53].

- *Internal barriers:* Legal departments tend to block collaboration as they regard the shareholder risk too high due to negative image when information about cyber vulnerabilities or incidents leaks through partners [53].

6. What's next?

6.1 Trends and developments in CIIP

CIIP is an ongoing challenge for governmental policymakers and political leadership. Effective CIIP requires a constant assessment of future technological developments and keeping track of the dynamics in the ICT and OT domains. The increasing use of ICT and (embedded) OT to monitor and control critical and complex cyber-physical systems means that most CI have CII components or are slowly transforming into CII. Meanwhile, the cyber security of OT is lagging far behind that of ICT despite specific cyber security good practices and standards [32, 55]. However, the IEC 62443 framework on Security for industrial automation and control systems has recently been extended with a part on RA [31].

Developments in ICT and OT and their interrelationships continuously alter the nature of CI and CII, for instance big data, smart energy grids, autonomously driving vehicles, 5G, e-health monitoring, and remote robotic surgery. Keeping track of the dynamically changing cyber risk landscape for CI and CII is therefore a challenge. Chapter 6 of [56] states that the "*continuous developments in digital technology require states to keep track of the changing risk landscape and to review CIIP policy accordingly*". Moreover, Chapter 4 of [11] states that "*Horizon scanning strengthens CIIP policy as it enables nations to proactively signal and assess developments in technology, and to act when new technology reaches the potential to become part of the national CII.*"

Nevertheless, it is difficult to recognize developments in the criticality of information infrastructures due to the hyper-connectivity of modern technologies which suddenly may alter existing dependencies and introduce new dependencies within CIIs and between CII and CI. Dependencies may shift in unforeseen ways due to unanticipated adoption of traditional or seemingly unimportant information infrastructure elements. Such changes may cause other information infrastructure services to become critical to a state on the one hand and to cause the criticality of other CII elements to disappear over time on the other hand [57].

Similarly, company policy changes unexpectedly may affect CI/CII incident response and recovery plans for ICT and OT operations. Consider the organization's green policy to replace all vehicles by e-vehicles. The existing incident response and recovery plans which dispatches repair trucks and their crews over long distances during a long power disruption will fail when no special provisions for recharging during non-normal modes of operation are made and will delay the recovery of CIs/CIIs.

Mass adoption and integration of new technologies such as internet of things (IoT), industrial internet of things (IIoT), internet-of-medical-things (IOMT), robotics and artificial intelligence may, besides changing the nature of CI and CII, also increase the risk of cyber and hybrid attacks to CII [34, 35]. Ecosystems of not well-secured, hundreds of thousands, if not more, of internetted devices may fall victim of cyber criminals. Their combined power may be used to attack CI, CII and life-essential devices, e.g. by denial of service attacks and spreading malware [58]. CI/CII operators and states shall be aware of this risk in time and take precautionary actions. For instance, smart grid technologies are fundamentally changing the energy sector and may introduce new CII elements at the national level. With the advancements in sensory, actuator and wireless technologies as well as the global internet, the usage of OT expands rapidly towards IIoT. The need for cyber security by design in new technological developments such as robotics and AI most often is an afterthought. This increases the cyber risk to CI, CII and humans, e.g. the use of robotic equipment such as vehicles and as human assistants in dangerous CI environments [59]. Moreover, new technologies enter the organization via the backdoor and is part of CI/CII services before the cyber risk is assessed and mitigated in a proper way.

6.2 Laws and regulations

The global cyber risk makes that states develop strategies, laws and regulations to get more grip on the cyber security risk to their state. Apart from the European general data protection regulation (GDPR) that became fully into effect in all EU Member States on May 25, 2018 [60], CI and some CII operators may be designated as operator of essential services (OES) or DSP as a result of the national law and related regulations which implement the EU NIS directive [13]. Whether one is designed as an OES or DSP depends on the service(s) provided, size of the operations, number of customers, area, and the level of criticality as laid down in national ruling. One requirement is that the OES or DSP shall notify the competent authority or the CSIRT with national authority without undue delay of any incident having a substantial impact on the provision of services. Moreover, national law may oblige notification by an OES to the 'CI stovepipe' responsible ministry or regulator. In case personal data is involved, the GDPR notification is required as well. Non-compliance with the law may result in a huge fine.

Reporting cyber incidents may lead to more transparency on the actual level of the cyber risk and may lead that to more awareness with operators and policymakers on the risk that cyber threats and vulnerabilities pose for society.

7. Conclusions

Analyzing the cyber risk in CI and CII, firstly requires the identification of CII using a set of (nationally) established criteria. RA for CI and CII may take place at multiple levels: by the organization of the CI/CII operator, by the CI/CII sector, nationally across all CI/CII sectors, and along the critical and essential service supply chains. This chapter provided insight to the OT risk, identifies the need for RA across organizations, and describes some RA models to address the cyber risk across multiple organizations and for service supply chains.

In assessing the cyber risk to CI/CII at the operator level, both ICT and OT should be considered. There exist many CI/CII sector-specific security control standards which can be mapped on common structures or frameworks as has been shown by e.g. NIST and ENISA. Although many standards and control measures exist, the OT risk at the governance, socio-technical, and operational-technical layers is often less understood and addressed by organizations. Recent advisories by government agencies show that the need to address the OT risk has become more urgent since the number of malicious attacks on OT as well as hybrid threats are growing while disruptions of the OT may have a large impact on the physical CI processes.

Recent research on RA for CI emphasizes on taking CI dependencies into account. This proves to be even more urgent and complex for CII. RA for CIIs and their dependencies is complex due to the highly dynamic nature of advances in and use of IT and OT, the often hidden nature of technological dependencies, think e.g.

about PNT services, and inclusion of embedded systems. Several RA approaches and methods exist to assess the cyber risk across organizations. However, assessing the cyber risk to the CI/CII service supply chains proves to be complex as it requires trust and willingness of all organizations involved.

And last not but least, organizations need to consider the cyber risk of future technologies before such technologies creep in via the backdoor and are an essential part of their critical services and business operations. The introduction of these new technologies can be planned (e.g. in the case of smart grids), which allows for an upfront analysis of the security risk involved, even when this risk is not always fully considered. New technologies, e.g. IoTs and dependencies may also be introduced in a more haphazard way into traditionally well-separated environments of CI/CII operators. Managing this additional risk is a major challenge for the operators.

Author details

Marieke Klaver[1]* and Eric Luiijf[2]

1 TNO Defence, Safety and Security, The Hague, The Netherlands

2 Luiijf Consultancy, Zoetermeer, The Netherlands

*Address all correspondence to: marieke.klaver@tno.nl

IntechOpen

© 2021 The Author(s). Licensee IntechOpen. This chapter is distributed under the terms of the Creative Commons Attribution License (http://creativecommons.org/licenses/by/3.0), which permits unrestricted use, distribution, and reproduction in any medium, provided the original work is properly cited.

References

[1] The Council of the European Union. Council Directive 2008/114/EC of 8 December 2008 on the identification and designation of European critical infrastructures and the assessment of the need to improve their protection. Off J Eur Union [Internet]. 2008;L345:75-82. Available from: https://eur-lex.europa.eu/LexUriServ/LexUriServ.do?uri=OJ:L:2008:345:0075:0082:EN:PDF

[2] CIPedia(c) [Internet]. [cited 2020 Sep 16]. Available from: http://www.cipedia.eu

[3] Ministry of Security and Justice. Working with scenarios, risk assessment and capabilities [Internet]. The Hague, Netherlands; 2009. Available from: http://www.itineris.nl/?mdocs-file=4987

[4] Pruyt E, Wijnmalen D. National Risk Assessment in The Netherlands. In: Ehrgott M, Naujoks B, Stewart TJ, Wallenius J, editors. Multiple Criteria Decision Making for Sustainable Energy and Transportation Systems [Internet]. Berlin, Heidelberg: Springer Berlin Heidelberg; 2010. p. 133-43. Available from: https://www.researchgate.net/publication/226282956_National_Risk_Assessment_in_The_Netherlands/stats

[5] Theocharidou M, Giannopoulos G. Risk assessment methodologies for critical infrastructure protection. Part II: A new approach [Internet]. Ispra, Italy; 2015. Available from: http://publications.jrc.ec.europa.eu/repository/bitstream/JRC96623/lbna27332enn.pdf

[6] Theocharidou M, Kotzanikolaou P, Gritzalis D. Risk-Based Criticality Analysis. In: Palmer C, Shenoi S, editors. Critical Infrastructure Protection III. Berlin, Heidelberg: Springer Berlin Heidelberg; 2009. p. 35-49.

[7] Critical Infrastructure Sector [Internet]. [cited 2020 Jul 23]. Available from: https://websites.fraunhofer.de/CIPedia/index.php/Critical_Infrastructure_Sector

[8] OECD. OECD Recommendation of the Council on the Protection of Critical Information Infrastructures Organisation for Economic Co-operation and Development C(2008)35. 2008.

[9] Boyes H, Isbell R. Code of Practice Cyber Security for Ships. London, United Kingdom; 2017.

[10] Colbert EJM, Kott A, editors. Cyber-security of SCADA and Other Industrial Control Systems [Internet]. Vol. 66. Boston, MA, USA: Springer; 2016. 354 p. Available from: http://link.springer.com/10.1007/978-3-319-32125-7

[11] Luiijf E, Van Schie T, Van Ruijven T. Companion Document to the GFCE-MERIDIAN Good Practice Guide on Critical Information Infrastructure Protection for governmental policy-makers. The Hague, The Netherlands; 2017.

[12] Stergiopoulos G. Power Sector Dependency On Time Service [Internet]. Heraklion, Greece; 2020. Available from: https://www.enisa.europa.eu/publications/power-sector-dependency/at_download/fullReport

[13] European Commission. Directive (EU) 2016/1148 of the European Parliament and of the Council of 6 July 2016 concerning measures for a high common level of security of network and information systems across the Union [Internet]. Brussels, Belgium; 2016. Available from: http://data.europa.eu/eli/dir/2016/1148/oj

[14] Council of the European Union. Regulation (EU) No 910/2014 of

the European Parliament and of the Council of 23 July 2014 on electronic identification and trust services for electronic transactions in the internal market and repealing Directive 1999/93/EC [Internet]. Vol. 57, Official Journal of the European Union. Brussels, Belgium; 2014. Available from: http://eur-lex.europa.eu/legal-content/EN/TXT/?uri=CELEX:32014R0910

[15] Cybersecurity and Infrastructure Security Agency. Information Technology Sector [Internet]. 2020 [cited 2020 Oct 26]. Available from: https://www.cisa.gov/information-technology-sector

[16] DHS. Commodification of Cyber Capabilities: A Grand Cyber Arms Bazaar [Internet]. Washington, DC, USA; 2019. Available from: https://www.dhs.gov/sites/default/files/publications/ia/ia_geopolitical-impact-cyber-threats-nation-state-actors.pdf

[17] De Bruijne M, Van Eeten M, Hernández Gañán C, Pieters W. Towards a new cyber threat actor typology: A hybrid method for the NCSC cyber security assessment [Internet]. The Hague, The Netherlands; 2017. Available from: https://www.wodc.nl/binaries/2740_Volledige_Tekst_tcm28-273243.pdf

[18] NSA, CISA. Cybersecurity Advisory NSA and CISA Recommend Immediate Actions to Reduce Exposure Across all Operational Technologies and Control Systems [Internet]. Washington, DC, USA: NSA and CISA; 2020. p. 1-5. Available from: https://media.defense.gov/2020/Jul/23/2002462846/-1/-1/1/OT_ADVISORY-DUAL-OFFICIAL-20200722.PDF

[19] NCSC. UPDATE: Vele Nederlandse Citrix-servers kwetsbaar voor aanvallen | Digital Trust Center [Internet]. The Hague, The Netherlands: National Cyber Security Centre; 2020. Available from: https://www.digitaltrustcenter.nl/nieuws/update-vele-nederlandse-citrix-servers-kwetsbaar-voor-aanvallen

[20] Budin G, Gréciano G, Rothkegel A, Hass U. Gestion du Risque pour usagers francophones [Internet]. Strasbourg, France; 2007. Available from: http://www.europhras.org/Site/anderedokumente/GMLGR5L_6_12_07.pdf

[21] Cabinet Office. Fact Sheet 2: National Security Risk Assessment [Internet]. London, United Kingdom; 2010. Available from: https://assets.publishing.service.gov.uk/government/uploads/system/uploads/attachment_data/file/62484/Factsheet2-National-Security-Risk-Assessment.pdf

[22] Trimintzios P, Gavrila R. National-level Risk Assessments [Internet]. Heraklion, Greece; 2013. Available from: https://www.enisa.europa.eu/publications/nlra-analysis-report/at_download/fullReport

[23] Public Safety Canada. All Hazards Risk Assessment Methodology Guidelines [Internet]. Ottawa, Canada; 2013. Available from: https://www.publicsafety.gc.ca/cnt/rsrcs/pblctns/ll-hzrds-ssssmnt/ll-hzrds-ssssmnt-eng.pdf

[24] European Commission. Regulation (EU) 2017/1938 of the European Parliament and of the Council of 25 October 2017 concerning measures to safeguard the security of gas supply and repealing Regulation (EU) No 994/2010 (Text with EEA relevance.). OJ L 280, 28102017 [Internet]. 2017;1-56. Available from: http://eur-lex.europa.eu/legal-content/EN/TXT/?uri=uriserv:OJ.L_.2017.280.01.0001.01.ENG&toc=OJ:L:2017:280:TOC

[25] NIST. Framework for Improving Critical Infrastructure Cybersecurity version 1.1 [Internet]. Gaithersburg, MD, USA; 2018. Available from: https://nvlpubs.nist.gov/nistpubs/CSWP/NIST.CSWP.04162018.pdf

[26] ISO. ISO 31000:2018 Preview Risk management -- Guidelines [Internet]. Geneva, Switzerland; 2018. Available from: https://www.iso.org/standard/65694.html

[27] ISO/IEC. ISO/IEC 27005:2018 Information technology — Security techniques — Information security risk management [Internet]. Geneva, Switzerland; 2018. Available from: https://www.iso.org/standard/75281.html

[28] ISO/IEC. ISO/IEC 27000:2018 Information technology — Security techniques — Information security management systems — Overview and vocabulary [Internet]. Geneva, Switzerland; 2018. Available from: https://www.iso.org/standard/73906.html

[29] ISO/IEC. ISO/IEC 27001:2013 Information technology — Security techniques — Information security management systems — Requirements [Internet]. Geneva, Switzerland; 2013. Available from: https://www.iso.org/standard/54534.html

[30] ISO/IEC. ISO/IEC 27002:2013 Information technology -- Security techniques -- Code of practice for information security controls [Internet]. Geneva, Switzerland; 2013. Available from: https://www.iso.org/standard/54533.html

[31] IEC. IEC 62443-3-2:2020: Security for industrial automation and control systems - Part 3-2: Security risk assessment for system design. Geneva, Switzerland; 2020.

[32] Stouffer K (NIST), Lightman S (NIST), Pillitteri V (NIST), Abrams M (MITRE), Hahn A (WSU). NIST Special Publication 800-82 Revision 2: Guide to Industrial Control Systems (ICS) Security [Internet]. Gaithersburg, MD, USA; 2015. Available from: https://nvlpubs.nist.gov/nistpubs/SpecialPublications/NIST.SP.800-82r2.pdf

[33] ENISA. Mapping of OES Security Requirements to Specific Sectors Mapping of OES Security Requirements to Specific Sectors About ENISA Mapping of OES Security Requirements to Specific Sectors [Internet]. Heraklion, Greece; 2017. Available from: www.enisa.europa.eu

[34] Niglia A, editor. Critical Infrastructure Protection Against Hybrid Warfare Security Related Challenges. Amsterdam, The Netherlands: IOS Press; 2020. 172 p.

[35] European Commission. Communication on the EU Security Strategy COM(2020) 605 final [Internet]. Vol. 53, Communication. Brussels, Belgium; 2019. Available from: https://ec.europa.eu/info/sites/info/files/communication-eu-security-union-strategy.pdf

[36] US-CERT. IR-ALERT-H-16-043-01AP CYBER-ATTACK AGAINST UKRAINIAN CRITICAL INFRASTRUCTURE [Internet]. Washington, DC, USA; 2016. Available from: https://ics-cert.us-cert.gov/alerts/IR-ALERT-H-16-056-01

[37] EURACOM. EURACOM Deliverable D2.3 - Integrated report on the link between Risk Assessment and Contingency Planning Methodologies [Internet]. Brussels, Belgium; 2010. Available from: https://docplayer.net/27429604-Deliverable-d2-3-integrated-report-on-the-link-between-risk-assessment-and-contingency-planning-methodologies.html

[38] Klaver MHA, Luiijf HAM, Nieuwenhuijs AH, Cavenne F, Ulisse A, Bridegeman G. European risk assessment methodology for critical infrastructures. In: Herder P, editor. 2008 First International Conference on Infrastructure Systems and

Services: Building Networks for a Brighter Future (INFRA) [Internet]. Rotterdam,The Netherlands: IEEE; 2008. p. 1-5. Available from: https://www.researchgate.net/publication/251883551_European_risk_assessment_methodology_for_critical_infrastructures

[39] Voster W, Mathijssen HH, Bloemen P, Beumer M, Dekker A, Mathijssen HH, et al. Cyber security supply chain risk analysis [Internet]. Dutch Cyber Security Council. The Hague, The Netherlands; 2015. Available from: https://www.cybersecurityraad.nl/binaries/Cybersecurity_supply_chain_risico-analyse_ENG_tcm107-323429.pdf

[40] ENISA. State-of-play of the implementation of the NIS Directive [Internet]. 2020 [cited 2020 Oct 30]. Available from: https://ec.europa.eu/digital-single-market/en/state-play-transposition-nis-directive

[41] Berg J van den, Zoggel J van, Snels M, Van Leeuwen M, Boeke S, Koppen L van de, et al. On (the Emergence of) Cyber Security Science and its Challenges for Cyber Security Education. In: NATO, editor. NATO STO/IST-122 symposium in Tallin, Estonia [Internet]. Paris, France: NATO RTA; 2014. p. 1-10. Available from: https://www.csacademy.nl/images/MP-IST-122-12-paper-published.pdf

[42] Luiijf E. Threats in Industrial Control Systems. In: Colbert EJM, Kott A, editors. Cyber-security of SCADA and Other Industrial Control Systems. Spinger; 2016. p. 69-94.

[43] RISI. Computer Sabotage at Nuclear Power Plant [Internet]. [cited 2018 Mar 3]. Available from: http://www.risidata.com/Database/Detail/computer_sabotage_at_nuclear_power_plant

[44] Lüders S. Control Systems under Attack? In: 10th ICALEPCS Int Conf on Accelerator & Large Expt Physics Control Systems Geneva, 10-14 Oct 2005, FR24-6O [Internet]. Geneva. Switzerland; 2005. p. 6. Available from: https://accelconf.web.cern.ch/accelconf/ica05/proceedings/pdf/O5_008.pdf

[45] Shodan [Internet]. [cited 2020 Sep 16]. Available from: https://www.shodan.io/

[46] UvW (Unie van Waterschappen). Baseline Informatiebeveiliging Waterschappen: Informatiebeveiliging Waterschappen Strategisch en Tactisch normenkader WS versie 1.0 [Internet]. Den Haag; 2013. Available from: https://www.uvw.nl/wp-content/uploads/2013/10/Baseline-Informatiebeveiliging-waterschappen-2013.pdf

[47] NERC. CIP-007-6 — Cyber Security – Systems Security Management Standard Development Timeline [Internet]. Washington DC, USA; 2014. (CIP). Available from: https://www.nerc.com/_layouts/PrintStandard.aspx?standardnumber=CIP-007-6&title=Cyber Security - System Security Management

[48] European Commission. Regulation (EU) 2019/881 of the European Parliament and of the Council on ENISA and on Information and Communications Technology Cybersecurity Certification and Repealing Regulation [Internet]. Official Journal of the European Union Brussels, Belgium; 2019 p. 15-69. Available from: https://eur-lex.europa.eu/legal-content/EN/TXT/PDF/?uri=CELEX:32019R0881&from=EN

[49] ENISA. Challenges of security certification in emerging ICT environments [Internet]. Heraklion, Greece; 2016. Available from: https://www.enisa.europa.eu/publications/challenges-of-security-certification-in-emerging-ict-environments

[50] ENISA. Considerations on ICT security certification in EU Survey Report [Internet]. Heraklion, Greece; 2017. Available from: https://www.enisa.europa.eu/publications/certification_survey

[51] JTF. Security and privacy controls for federal information systems and organizations Rev 5. Vol. 800, NIST Special Publication. Gaithersburg, MD, USA; 2020.

[52] ENISA. Cyber Security Information Sharing: An Overview of Regulatory and Non-regulatory Approaches [Internet]. Heraklion, Greece; 2015. Available from: https://www.enisa.europa.eu/publications/cybersecurity-information-sharing/at_download/fullReport

[53] Luiijf E, Kernkamp A. Sharing Cyber Security Information, Good Practice Stemming from the Dutch Public-Private-Participation Approach, The Hague, Netherlands. 2015.

[54] Klaver MHA, Vos P, Tjemkes B, Verner DR. Enhancement of Public-Private Partnerships within Critical Infrastructure Protection Programs. The Hague, Netherlands; 2017.

[55] DHS (Department of Homeland Security). Recommended Practice: Improving Industrial Control Systems Cybersecurity with Defense-In-Depth Strategies [Internet]. Washington, DC, USA; 2016. Available from: https://ics-cert.us-cert.gov/sites/default/files/recommended_practices/NCCIC_ICS-CERT_Defense_in_Depth_2016_S508C.pdf

[56] Luiijf, H., van Schie T, van Ruijven T, Huistra A. The GFCE-MERIDIAN Good Practice Guide on Critical Information Infrastructure Protection for governmental policy-makers. The Hague, The Netherlands: TNO; 2016.

[57] Luiijf E, Klaver M. Governing critical ICT: Elements that require attention. Eur J Risk Regul. 2015;6(2):263-70.

[58] De Donno M, Dragoni N, Giaretta A, Spognardi A. DDoS-Capable IoT Malwares: Comparative Analysis and Mirai Investigation. Bugliesi M, editor. Secur Commun Networks [Internet]. 2018;2018:7178164. Available from: https://doi.org/10.1155/2018/7178164

[59] Steijn W, Luiijf E, Beek D van der. Emergent risk to workplace safety as a result of the use of robots in the work place [Internet]. Utrecht, The Netherlands; 2016. Available from: http://publications.tno.nl/publication/34622295/QDXZqU/steijn-2016-emergent.pdf

[60] European Commission. Regulation (EU) 2016/679 of the European Parliament and of the Council of 27 April 2016 on the protection of natural persons with regard to the processing of personal data and on the free movement of such data, and repealing Directive 95/46/EC (General Da [Internet]. Brussels, Belgium; 2016. Available from: http://eur-lex.europa.eu/legal-content/en/TXT/?uri=CELEX%3A32016R0679

Chapter 10

Flood Risk Analysis for Critical Infrastructure Protection: Issues and Opportunities in Less Developed Societies

Ugonna C. Nkwunonwo

Abstract

This chapter presents all-important discussions relating to flood risk analysis which arguably is a subject of overwhelming significance within the context of less developed societies, for example Nigeria. Whilst a possible means of eradicating flooding from human environment is inconceivable, debates for more effective flood risk reduction methodologies for critical infrastructure protection must continue. Increased population and urbanisation scenarios drive worsened flood risk which trigger increased efforts for corporate adaptability to flooding. To ensure that social systems can cope with floods, it is important to investigate why best practices in flood risk reduction are not fully applicable. This chapter explores these issues drawing from extant dialogues on flood risk management (FRM). Arguably, the current flood modelling techniques and assessment of vulnerability operations largely do not support a realistic analysis of flood risk. Funnelled through an interpretative research paradigm, the chapter conceives that these limitations fall under five cardinal issues – (1) data, (2) theories and concepts, (3) existing flood risk analyses methods, (4) legislation and policy, and (5) sustainable development. It argues that the realisation of a more effective flood risk reduction for the poorer and less developed societies will depend on effective tackling of these issues which creates opportunities for flood risk analyses through simplified approaches, and use of free and open geospatial data infrastructure.

Keywords: flooding, flood risk analyses, less developed societies, urbanisation, flood risk management, flood modelling, vulnerability assessment

1. Introduction

The widespread flooding in recent times, the reduction of its impacts on human populations, properties and economic activities and the impracticability of its eradication from natural environment are factors of global concerns [1–4]. In the developing countries (DCs) such as Nigeria, there is evidence to suggest that the thought of the next flooding event appears to apprehend many local communities, urban residents and authorities' hierarchy [5]. Arguably, this reality suggests among other interesting discourse, that the recognition of flooding impacts and the curiosity they drive in human populations are fundamental towards finding realistic solutions to the hazard. Worthy of note within this context is the damage potential of flooding which is debatably unprecedented when compared to other known environmental hazards occurring within natural human environment in recent times [6, 7]. Whilst the failure and/or limitation of efforts to tackle flooding are issues in the DCs that clearly require urgent attention [8, 9], the need for sustainable development which underpins adaptability and collective resilience of the general public to flooding cannot be disregarded [10, 11].

The interplay between causes, impacts and remedies of flooding phenomenon highlights the situation in the less developed societies in respect of flooding and the risk it poses. This chapter focuses on the Lagos metropolis of Nigeria in West Africa. Under the quandary of rapid population increase and urbanization, it appears the conurbation has been subject to critical and disturbing scenarios. The idea that population growth will compel worsened future flood risk highlights the need to engage with more proactive measures of tackling flooding and more importantly more effective means of building the capacities of human population to cope with floods [11, 12]. However, present efforts at addressing the challenges of flooding in the Lagos area are flawed [13, 14]. Whilst the area signifies the economic and industrial hub of Nigeria and attracts tourists from within the country and abroad, responses to security challenges, poor corporate adaptation and resilience to flooding among other besetting environmental hazards is inadequate [15, 16]. Existing knowledge regarding particularly to the state of affairs of flooding in Lagos is unsatisfactory and falls short of solutions to the impacts of the hazard on human populations and has been unable to support sustainable development within the region [14, 17].

Within these contexts, it is imperative that the critical factors which undermine efforts at tackling flooding in Lagos as well as gaps in knowledge among other considerations which can be associated with increasing flood risk generally are identified. Thus, the need to support present efforts at tackling flooding in the Lagos region and to advance existing knowledge relating to flood risk reduction in the area motivate the debates in this chapter which considers a triplet of objectives: firstly, to summarise the widespread flooding in the Lagos metropolis of Nigeria, secondly, to summarise the current efforts towards tackling the hazard in Lagos and to identify key limitations and gaps in knowledge and practice, and finally, whilst the author argues that inadequate flood modelling in the area and limited application and scope of assessment of vulnerability to flooding undermine the success of current efforts to tackling flooding, why more of such investigations are needed is presented along with the possible challenges facing their applications in Lagos, Nigeria. It equally presents the prospects for flood risk analyses through simplified approaches and open geospatial data.

2. Widespread flooding in Lagos Nigeria

Past and present flooding in Lagos Nigeria, highlight the influence of climate change, rapid population growth and urbanization on the local hydrology of the region [18–20]. First and foremost, the Lagos metropolis consists of 16 local government areas (LGAs) of varying spatial enumeration units (the largest being about 194 km^2) (see **Table 1** and **Figure 1**). The total land mass of the conurbation exceeds 1100 km^2. Based on the state government's statistics [22], up to 21 million people reside in the area and this creates a yawning dimension of adverse social and environmental condition mostly overcrowding and slum development. The lack of space for a myriad of anthropogenic activities forces development of flood prone areas thus instigating a severe vulnerability to flooding for those inhabitants who lack social capacities to cope with the hazard. The abundance of impervious surfaces

S/no.	LGAs	Land area (km^2)
1	Agege	11.263
2	Ajeromi-Ifeledun	12.395
3	Alimosho	186.195
4	Amuwo-Odofin	135.240
5	Apapa	26.798
6	Eti-osa	193.460
7	Ifako-Ijaiye	26.769
8	Ikeja	46.427
9	Kosofe	81.889
10	Logos-island	8.707
11	Lagos-mainland	19.572
12	Mushin	17.576
13	Ojo	158.884
14	Oshodi-Isolo	44.999
15	Shomolu	11.615
16	Surulere	23.122

Table 1.
16 local government areas and their spatial units in the Lagos metropolis of Nigeria. Source: Adapted from [21].

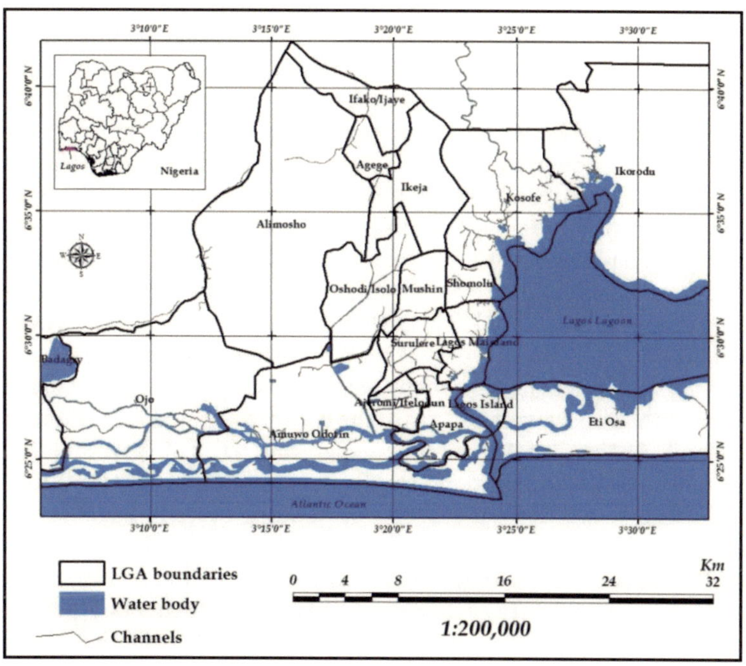

Figure 1.
The Lagos metropolis of Nigeria. Inset showing the location of Lagos State in Nigeria. Source: Drafted by authors.

in the area which generally causes increased surface water runoff and reduced soil infiltration highlights the impediments of poor urban drainage system [23].

Following the overview of Lagos metropolis presented in the preceding paragraph, a clearer picture of the devastating effects of flooding in the area can be

Figure 2.
Examples of flooding scenes in the Lagos metropolis of Nigeria: (a) living room submerged by flood water, (b) residential building submerged, (c) local community affected by flood waters, and (d) expressway overwhelmed by flood water. Source: Online images of flooding in Lagos, Nigeria.

appreciated. Typically, the hazard which has been generally attributed to climate change and poor urban planning affects hundreds of thousands of people (mostly through homelessness, physical injuries, mortality, spread of diseases and emotional trauma), destroys chains of urban infrastructure and disrupts economic activities [14, 24]. Fiscal losses caused by the hazard in the area amount to millions of US dollars [25]. Although fluvial and coastal flooding occurred in the early days of flooding – i.e. in the early 1960s – pluvial floods resulting from prolonged rainfall which overwhelms urban drainage facilities and soil infiltration capacity are now more widespread. Such floods usually recur annually between the months of March and October (but usually more severe in July) with considerable environmental and socio-economic impacts [26]. These floods which have triggered concerns for environmental mismanagement, urgent humanitarian needs and services, primary health delivery, solid waste management, urban development and governance, and the resilience of the general public within the area are claimed to be more severe for Lagos Island, Apapa, Ikeja, Mushin, Surulere and parts of Ikorodu [16, 26–29]. The magnitude of flooding experience in the Lagos region of Nigeria is highlighted in **Figure 2(a-d)** below. A typical example of flooding event in Lagos is the July 2011 flooding, caused by a heavy rainfall that lasted 17 hours. The flood affected more than 10 thousand people with deaths exceeding 100 and a range of damage including public infrastructure such as roads, bridges and schools. Houses were submerged by flood water while lots of properties including vehicles were destroyed due to the intensity of the flood. An estimated economic loss of about 50 billion Naira ($US 320 million) was incurred [30].

3. Summary of current efforts towards tackling flooding in Lagos

For appraisal of current approaches to flood risk management in Lagos city by public and private sectors and the implications of such approaches within the context of global flood risk management practices, refer to Adelekan [31]. Flooding and the means of tackling its risk in the Lagos metropolis of Nigeria have received considerable attention in the literature since the last two decades, and this arguably demonstrates commitment on the part of the Lagos state government and various stake holders. Some of the ongoing practices as argued by Oshodi [32] include: expansion of drainage infrastructure within the city heartland, annual debris removal from principal drainage facilities within the city heartland, recommendation and resettlement of the dwellers of flood plains and wetlands and the residents of Ogun river catchment areas, demolition of homes in the flood prone areas. Several flood preventive and curative initiatives ranging from community self-assistance actions to World Bank assisted programmes were identified in Odunuga [26]. Recently, key initiatives which include the Drain Dock and The Emergency Flood Abatement Gang (EFAG) were launched by the government of Lagos state to improve current efforts towards addressing the challenges of flooding. Action by the ministries of Environment, Works and Health as well as the Lagos Metropolitan Development and Governance Project (LMDGP) towards controlling flood hazard in the area including waste management programme, shoreline protection, low carbon emission, school advocacy programme and climate change club are acknowledged [14].

It can be shown that how to promote sustainable drainage infrastructure and sustainable access to basic urban services for urban residents and the general public are a top priority. Lagos state emerged as the first in Nigeria to carry out a detailed topographic mapping of the area with LiDAR (Light detection and ranging) data acquisition and GIS based analysis aimed at addressing the challenges of flooding. Although it is claimed that these measures have been preventive in context, they are unprecedented in Nigeria and clearly demonstrate practical commitment to fighting flooding [33, 34]. However, in the light of 'best practices' in flood risk reduction and 'lessons learned' from other countries' experiences of flooding, it can be argued that such measures are at best limited. Although the lack of data, lack of funds and improved technology as well as poor political will have been implicated [23, 35], flood modelling which is needed to systematically tackle flooding within the context of flood risk/hazard mapping and provision of flood data for improving the perception of flooding among the general public and to support other non-structural approaches to flood risk reduction seems to have been ignored.

4. Flood modelling and assessment of vulnerability to flooding for the Lagos metropolis of Nigeria

Flood risk reduction is fundamentally a knowledge-driven ideology that shapes the pathway towards living with floods. Key knowledge that drives this idea is often based on flood risk/hazard maps, public opinions and specialist judgement on flooding. Within this perspective, flood modelling which predicts flood data (mainly flood water depth, duration and extent, as well as depth-averaged velocities) essentially needed for flood risk/hazard analyses, mapping and assessment plays significant roles [36, 37]. Conceptually, flood modelling may be perceived as a scientific technique that numerically or analytically solves relevant governing mathematical equations and generates computer algorithms and codes for fast, continuous and routine simulation of flood data [9]. Quick, continuous and routine provisions of flood data appear to undermine ground survey methods and remote sensing technologies, thus most evidently highlighting the relevance of flood modelling.

For the Lagos area, besides the importance of quick, continuous and routine provision of flood data, it is pertinent to realize the specific roles which flood modelling can play towards flood risk reduction and these includes: (1) to align the goals of flood risk management in the Lagos areas with the objectives of such roles in other places such as the United States, United Kingdom and the Netherlands, (2) to pave the way for overcoming the various hassles associated with flood modelling generally such as computation complexity and model instability/conditional stability, (3) to strengthen the means of improving flood awareness among urban residents and other stake holders through flood risk/hazard mapping, and (4) to combine with vulnerability assessment in order to build the capacity of a wider population to cope with floods.

Vulnerability is clearly a relevant concept in disaster/risk management and it suggests the propensity to which a system, subsystem or systems component can be adversely affected by a stressor [38]. System, subsystem or systems component refer to human populations and/or critical infrastructure that appear to be in harm's way during flood hazard occurrence which is the reason why exposure, sensitivity and adaptive capacity are often considered in the course of analysing vulnerability [38, 39]. In the Lagos metropolis of Nigeria, the issue of vulnerability to flooding is critical given that urbanization and rapid population growth which both trigger and increase slum development and development on flood prone areas [40]. However, this odd scenario has not been sufficiently tackled with adequate knowledge of vulnerabilities of social systems to flooding and the factors that influence such vulnerabilities. Few studies that considered vulnerability to flooding in Lagos are limited in scope, constrained by paucity of quality data and narrowed discussions down to small areas [15, 16, 41, 42]. It can be shown that results obtained from analysing vulnerability to flooding at such small scales cannot be generalized for the Lagos area [43].

5. The challenges of flood modelling and vulnerability assessment in the Lagos metropolis of Nigeria

Given the general merits of flood modelling and assessment of vulnerability to flooding and the specific roles they can play towards flood risk reduction in the Lagos metropolis of Nigeria (refer to [42]), it is important to identify the factors that potentially constrain their application. In view of this, the author conceived and discuss the following issues:

5.1 Issues on data

The fact that flood modelling and assessment of vulnerability to flooding require sufficient and accurate data to implement suggests paucity of data as mostly constraining such operations. For the Lagos area, it can be argued that issues relating to relevant data can be likened to a total mirage ranging from abject paucity, inaccuracy and limited access. A typical example is demography for which two key sources (2006 National Population Census and Lagos State Digest of Statistics) quoted different figures representing the Lagos region. Equally complicating is media reporting which has been inconsistent in many instances [44]. Although high resolution LiDAR data is now available for the area, access to the dataset for flood modelling and assessment of vulnerability is been constrained by cost.

5.2 Issues on theories and concepts

Flood risk reduction is a key concern for major environmental research themes (for example Climate Change Adaptation (CCA) and Disaster Risk Reduction (DRA)) which promote the development of integrated methodologies subject to living with floods rather than fighting them [45, 46]. Invariably, such methodologies seem to require in-depth understanding of the drivers of flood risk while their practicability appears to suit ideal situation favoured by easy access to relevant datasets and technical requirements. However, these methodologies often lack sufficient flexibilities for application to external case studies such as the DCs. To circumvent such methodology inflexibility, it is imperative that new methodologies are developed. For the Lagos area, it can be argued that the development of new methodologies with sufficient capacity to support assessment of vulnerability to flooding and flood modelling can be easily undermined by the underlying concepts and theories which are generally inductive based on ontological perspective.

5.3 Issues on existing methodologies

Expectations are increasing for more efficient methodologies with regards to tackling flooding and the risks it poses [47]. Based on this, improving on the functionality of existing methodology has become a popular hypothesis recently. Whilst this assumption has been affirmed in many cases, intuitively, an important concept such as flood modelling underlines the need to understand the basic components that limits existing methodologies [48, 49]. Within the context of flood modelling, existing methodologies (especially for the physically based numerical flood models) lead to models that are computationally expensive, often unstable/conditionally stable requiring a certain CFL condition (Courant-Freidrichs-Lewy condition), which prescribes small time steps leading to high computation burden. Besides, some of these models lack extensive calibration due to insensitivity to certain parameterisation. For the Lagos metropolis of Nigeria, it is argued that the means to overcome these challenges present a critical consideration which undermines flood modelling in the area, although the Lack of funds to acquire commercial codes along with their technical assistance can also have a resistive impact on flood modelling [9].

5.4 Issues on legislation and policy

Flood risk reduction within the context of living with floods is strengthened by robust legislation towards environmental management, intensive research and adaptation of human population to the hazard. Nigeria among many DCs is characterized by weak legislation towards hazard management [50]. This arguably impacts negatively on the inclusion of more robust approaches such as flood modelling and assessment of vulnerability to tackling flooding and the challenges it poses. As argued by Oshodi [32], due to the weak legislation and poorly implemented policies regarding hazard risk and environmental sustainability in the Lagos area, full preparedness to deal with the challenges of flooding is uncertain.

5.5 Issues on sustainable development

Flood modelling and assessment of vulnerability to flooding required to effectively tackle flooding underpin sustainable development [11]. Within the context of sustainable development, every society aspires towards meeting human development goals while sustaining the ability of natural systems to continue to provide the natural resources and ecosystem services upon which the economy and society depend [51, 52]. Despite much attention which it has received, sustainable development in the DCs remains uncertain and almost unrealistic due to a number of factors for examples: gender inequality, poverty, weak legislative impetus, governance and political will, sluggish judicial administration and access to justice, corruption, asymmetric corporate social irresponsibility and poor access to information, and technical knowledge [53, 54]. For the Lagos area of Nigeria, poor public participation in planning, capacity building, and integration of information technology into planning practice are key factors that constrain sustainable development [55]. Poor public participation can be revealed mainly in the poor awareness of flooding among the wider public, and lack of compliance to environmental laws. To investigate the vulnerability to flooding of social systems for example, relevant information is often derived from public survey and responses to questionnaire. Arguably, inaccurate or uncorrelated responses from questionnaires which jeopardize the outcomes of such investigations can result from poor awareness of flooding.

6. Opportunities for flood risk analyses through simplified approaches and free and open geospatial data in the less developed societies

Kovacs *et al.* [56] compiled a French technical report of several simplified approaches to flood risk analyses in the developing countries. These techniques are simplified in theory and often require utilise freely available datasets for flood risk analyses and protection of critical infrastructure in the less developed societies. Several other attempts have been made in the literature. The prospects within these simple techniques to enable stakeholders lessen the threats of flooding on critical infrastructure and sustainable development are significant. Hammond *et al.* [57] developed a modified Drivers-Pressures-State-Impact-Response (DPSIR) framework which enables policy makers to evaluate strategies for improving flood resilience in cities. Nkwunonwo *et al.* [49] proposed the new scheme, *GFSP-1,* to model urban flooding using a minimum of data. The model which was implemented in a MATLAB environment was tested using the flooding event of year 2000 in Portsmouth, UK, and later used to simulated the historic flooding of year 2011 in the Lagos area of Nigeria. See *et al.* [58] utilised an open data approach which includes open street map and field paper to map urban drainage infrastructure in the Philippines. Results emerging from these simplified approaches correlate positively with real life data, and have been effective in assessing flood risk and vulnerabilities, and providing realistic feedbacks to stake holders.

The major weakness in these simplified approaches is the lead time in moving towards an integrated flood risk management. This is because of many assumptions made to actualise data fitting in the simple methods, and the inability of the simple techniques to capture all the physical parameters and nexus around the variables that motivate flooding within catchment area. This increases epistemic and aleatory uncertainty, and makes it hard to generalise the methods towards a more effective stimulus in flood risk management. Flood risk is an aggregate of multiple factors – hazard, exposure and vulnerability – drawing from Crichton's risk triangle [59]. Land use analyses and flood modelling are able to evaluate the magnitude of exposure and flood hazard (depth and extent along with velocity of flood water) [9, 60, 61]. Vulnerability is a bit more practical because of its conceptualization and theories that underpin its analyses. In the current literature, flood vulnerability is a measure of elements at risk of flooding because they lack coping capacity or any form of adaptive mechanism. It is an ideal science culture to includes community participation in analysing flood vulnerability. This is standard technique in the developed societies, and few authors have discussed its application in the Lagos area of Nigeria. Although, data paucity and challenges adapting existing methodologies to new case studies often stand in the way of an ideal vulnerability analyses, participatory approaches for collecting informal knowledge on exposed elements and vulnerabilities from the population and local actors is invaluable towards assessment of vulnerability to flooding. Douglas *et al.* [19] Adelekan [13, 15], and Salami *et al.* [62] used this approach in the Lagos area of Nigeria, and the result have been fundamental to decision support in flood risk management within the area.

One clear insight into stare-of art methodologies for vulnerability assessment is the importance of indicators as proxies to vulnerability variables. Several studies have applied this method for examples Müller *et al.* [63] and Tapia *et al.* [64] and the outcomes demonstrate how the various types of vulnerability – physical, economic and social – not only relate to various dimensions of the society, but also varies according to the complexity and main determinants of sustainable development. This understanding plays important roles in protecting critical infrastructure from flooding, bearing in mind the question of what makes a critical infrastructure vulnerable? For example, social vulnerability is based on social factor such as age, gender, socio-economic status [65], an idea which Nkwunonwo [41] applied, using demographic distribution from Nigeria's 2006 census to assess the social vulnerability of Lagos to urban flooding. Indicator-based vulnerability analyses is often complicated by the lack of method to measure a particular indicator. In such a situation, expert elicitation based on observed vulnerabilities and impacts following a previous catastrophic event can been used for vulnerability assessment and modelling. This is a simple approach that serves the purpose and addresses the gap in flood risk assessment in developing societies.

7. Conclusion

Flooding experiences in the Lagos metropolis of Nigeria are overwhelming and has remained an issue of incessant debate. Although there are present efforts at tackling the hazard, success so far has arguably been limited and ample discussion regarding this condition are critical. Whilst flooding is generally accepted as an inevitable phenomenon in present day environment, reducing its impacts on people and the environment is a significant priority for many regional and international flood management initiatives and directives [3, 66]. To achieve the sole aim of flood risk reduction which is "living with floods rather than fighting them", flood modelling and assessment of vulnerability to flooding are fundamental operations and have been applied in many developed countries such as the United States, United Kingdom and Netherlands [67]. However, for the Lagos metropolis of Nigeria, flood modelling and assessment of vulnerability to flooding have been skimped.

As a critical focus, this chapter makes attempts to bridge the gaps in knowledge and practice of flood risk reduction in the Lagos area and investigates the key reason why these approaches were skimped in the Lagos area. It is argued that unless these critical issues such as limitation in data, legislation and policy and mismatch in sustainable development, the application of flood modelling and assessment of vulnerability to flooding in the Lagos metropolis of Nigeria will remain unrealistic. Moreover, simplified approaches and freely available and open source datasets create opportunities to undertake flood risk assessment despite the issues that cause severe limitations. Research is needed to provide bespoke methodologies that will take advantage of these resources to provide workable feedbacks to stake holders and flood risk management policy males.

Author details

Ugonna C. Nkwunonwo
Department of Geoinformatics and Surveying, University of Nigeria, Enugu Campus, Nigeria

*Address all correspondence to: ugonna.nkwunonwo@unn.edu.ng

IntechOpen

© 2020 The Author(s). Licensee IntechOpen. This chapter is distributed under the terms of the Creative Commons Attribution License (http://creativecommons.org/licenses/by/3.0), which permits unrestricted use, distribution, and reproduction in any medium, provided the original work is properly cited.

References

[1] Kirezci, E., Young, I. R., Ranasinghe, R., Muis, S., Nicholls, R. J., Lincke, D., & Hinkel, J. (2020). Projections of global-scale extreme sea levels and resulting episodic coastal flooding over the 21st Century. Scientific Reports, *10*(1), 1-12.

[2] Montz, B. E. (2020). Risk management: Are there parallels between COVID19 and floods?. Journal of Flood Risk Management, *13*(2), 1-10.

[3] Sayers P, Li Y, Galloway G, Penning-Rowsell E, Shen F, Wen K, et al. *Flood Risk Management: A Strategic Approach*. Paris: UNESCO; 2013

[4] Schober, B., Hauer, C., & Habersack, H. (2020). Floodplain losses and increasing flood risk in the context of recent historic land use changes and settlement developments: Austrian case studies. *Journal of Flood Risk Management*, e12610.

[5] Ferreira, S., & Ghimire, R. (2012). Forest cover, socioeconomics, and reported flood frequency in developing countries. Water Resources Research, *48*(8), 1-10.

[6] Kellermann, P., Schröter, K., Thieken, A. H., Haubrock, S. N., & Kreibich, H. (2020). The object-specific flood damage database HOWAS 21. Natural Hazards and Earth System Sciences, *20*(9), 2503-2519.

[7] Tay, C. W., Yun, S. H., Chin, S. T., Bhardwaj, A., Jung, J., & Hill, E. M. (2020). Rapid flood and damage mapping using synthetic aperture radar in response to Typhoon Hagibis, Japan. Scientific Data, *7*(1), 1-9.

[8] Lumbroso, D. (2020). Flood risk management in Africa. Journal of Flood Risk Management, *13*(3), 1-9.

[9] Nkwunonwo, U. C., Whitworth, M., & Baily, B. (2020). A review of the current status of flood modelling for urban flood risk management in the developing countries. Scientific African, *7*, e00269.

[10] Chan, F. K. S., Griffiths, J. A., Higgitt, D., Xu, S., Zhu, F., Tang, Y. T., Yuyao, X., & Thorne, C. R. (2018). "Sponge City" in China—a breakthrough of planning and flood risk management in the urban context. Land Use Policy, *76*, 772-778.

[11] Juarez Lucas, A. M., & Kibler, K. M. (2016). Integrated Flood Management in developing countries: balancing flood risk, sustainable livelihoods, and ecosystem services. International Journal of River Basin Management, *14*(1), 19-31.

[12] Abbas, A., Amjath-Babu, T. S., Kächele, H., Usman, M., & Müller, K. (2016). An overview of flood mitigation strategy and research support in South Asia: implications for sustainable flood risk management. International Journal of Sustainable Development & World Ecology, *23*(1), 98-111.

[13] Adelekan, I. O. (2016). Flood risk management in the coastal city of Lagos, Nigeria. Journal of Flood Risk Management, *9*(3), 255-264.

[14] Nkwunonwo, U., Whitworth, M., & Baily, B. (2016). A review and critical analysis of the efforts towards urban flood risk management in the Lagos region of Nigeria. Natural Hazards and Earth System Sciences, *16*(2), 349-369.

[15] Adelekan, O. (2010). Vulnerability of poor urban coastal communities to flooding in Lagos, Nigeria. Environment and Urbanization, 22, 433-450.

[16] Ajibade, I., McBean, G., & Bezner-Kerr, R. (2013). Urban flooding in Lagos, Nigeria: Patterns of vulnerability

and resilience among women. Global Environmental Change, *23*, 1714-1725.

[17] Adelekan, I. O., & Asiyanbi, A. P. (2016). Flood risk perception in flood-affected communities in Lagos, Nigeria. Natural Hazards, *80*(1), 445-469.

[18] ActionAid. *Climate change, urban flooding and the rights of the urban poor in Africa: Key findings from six African cities*. London: Action Aid International; 2006

[19] Douglas, I., Alam, K., Maghenda, M., Mcdonnell, Y., McLean, L., & Campbell, J. (2008). Unjust waters: climate change, flooding and the urban poor in Africa. Environment and Urbanization, 20(1), 187-205.

[20] Elias, P., & Omojola, A. (2015). The challenges of climate change for Lagos, Nigeria. Current Opinion in Environmental Sustainability, *13*, 74-78.

[21] National Population Commission (NPC) (2006). 2006 Population and Housing Census: National and State Population and Housing Tables: Priority Tables I-IV. FCT, Abuja: Federal Government of Nigeria.

[22] Lagos State Government (LSG). Abstract of Local Government Statistics. Lagos: Lagos Bureau of Statistics, Ministry of Economic Planning and Budget Secretariat, Alausa, Ikeja; 2012

[23] Adeloye, A.J., & Rustum, R. (2011). Lagos (Nigeria) flooding and influence of urban planning. *Urban Design and Planning*, 164 (DP3), 175-187.

[24] Aderogba, K. A. (2012). Global warming and challenges of floods in Lagos metropolis, Nigeria. Academic Research International, 2(1), 448-468.

[25] EM-DAT (The international Disaster Database) (2014). Centre for Research on the Epidemiology of Disasters - CRED. Flooding data for Nigeria. 2014; Accessed 10th March 2015. Available at: www.emdat.be/

[26] Odunuga S. (2008). Urban land use change and the flooding in Ashimowu watershed, Lagos, Nigeria. University of Lagos, Nigeria: PhD thesis.

[27] Lamond, J., Stanton-Geddes, Z., Bloch, R., & Proverbs, D. (2013). *Cities and Flooding: Lessons in resilience from case studies of integrated urban flood risk management*. CIB.

[28] Oyebande L. *Drainage protection to urban lands: an environmental challenge*. Nigerian Geographical Association Conference: University of Nigeria, Nsukka, Enugu; 1974

[29] Soneye, A. (2014). An overview of humanitarian relief supply chains for victims of perennial flood disasters in Lagos, Nigeria (2010-2012). Journal of Humanitarian Logistics and Supply Chain Management, 4(2), 179-197.

[30] Oladunjoye, M. (2011). *Nigeria: July 10 Flooding – Lagos Gives Relief Materials to Victims*. Daily Champion Newspaper. http://allafrica.com/ stories/ 201109080792.html (accessed 08/02/2015).

[31] Adelekan, I. O. (2015). Flood risk management in the coastal city of Lagos, Nigeria. *Journal of Flood Risk Management*. DOI: 10.1111/jfr3.12179

[32] Oshodi, L. (2013). Flood management and governance structure in Lagos, Nigeria. Regions Magazine, *292*, 22-24. DOI: 10.1080/13673882.2013.10815622.

[33] Njoku, J.D., & Udeagha, M. (2013). Assessing the flooding potentials of Oguta lake Watershed using Remote Sensing Technology. Paper presented at 5th Annual National Conference organized by the Nigerian Association of Hydrological

Sciences (NAHS) at University of Nigeria, Nsukka. 21-30.

[34] Obeta, C. M. (2014). Institutional Approach to Flood Disaster Management in Nigeria: Need for a Preparedness Plan. British Journal of Applied Science & Technology, 4 (33), 4575- 4590.

[35] Nkwunonwo, U. C., Whitworth, M., Baily, B., & Inkpen, R. (2014). The development of a simplified model for urban flood risk mitigation in developing countries. In Vulnerability, Uncertainty, and Risk@ Quantification, Mitigation, and Management (pp. 1116-1127). ASCE.

[36] Cai, T., Li, X., Ding, X., Wang, J., & Zhan, J. (2019). Flood risk assessment based on hydrodynamic model and fuzzy comprehensive evaluation with GIS technique. International Journal of Disaster Risk Reduction, *35*, 101077.

[37] Tsakiris, G. (2014). Flood risk assessment: concepts, modelling, applications. Natural Hazards and Earth System Sciences, *14*(5), 1361.

[38] Adger, W. N. (2006). Vulnerability. Global Environmental Change, 16(3), 268-281.

[39] Turner II., B.L., Kasperson, R.E., Matson, P.A., McCarthy, J.J., Corell, R.W., Christensen, L., Eckley, N., Kasperson, J.X., Luers, A., Martello, M.L., Polsky, C., Pulsipher, A., Schiller, A., (2003). A framework for vulnerability analysis in sustainability science. Proceedings of the National Academy of Sciences US 100, 8074-8079

[40] Agbola, T. and Agunbiade, E. (2007). Urbanization, slum development and security of tenure: the challenges of meeting millennium development goal (MDG 7) in metropolitan Lagos, Nigeria. In PRIPODE Workshop, Nairobi, Kenya (pp. 11-13).

[41] Nkwunonwo, U. C. (2017). Assessment of social vulnerability for efficient management of urban pluvial flooding in the Lagos metropolis of Nigeria. Journal of Environmental Studies, 3, 1-11.

[42] Nkwunonwo, U. C., Whitworth, M., & Baily, B. (2015). Relevance of social vulnerability assessment to flood risk reduction in the Lagos metropolis of Nigeria. British Journal of Applied Science & Technology, 8(4), 366-382.

[43] Tapsell, S., McCarthy, S., Faulkner, H., & Alexander, M. (2010). Social vulnerability to natural hazards. *State of the art report from CapHaz-Net's WP4*. London.

[44] Olalekan, A. (2013). The Inconsistency of the Flood Narrative in Nigeria. http://www.e-ir.info/2013/02/04/the-inconsistency-of-the-flood-narrative-in-nigeria/

[45] Di Baldassarre, G., & Uhlenbrook, S. (2012). Is the current flood of data enough? A treatise on research needs for the improvement of flood modelling. Hydrological Processes 26, 153-158.

[46] UN/ISDR: United Nations International Strategy for Disaster Reduction. (2004). Living with Risks: A global Review of Disaster Reduction Initiatives. 2004 Version Volume 1. http://www.unisdr.org/files/657_lwr1.pdf

[47] Demeritt, D., & Nobert, S. (2014). Models of best practice in flood risk communication and management. Environmental Hazards, 13(4), 313-328, DOI:10.1080/17477891.2014.924897

[48] Hunter, N. M., Bates, P. D., Horritt, M. S., & Wilson, M. D. (2007). Simple spatially-distributed models for predicting flood inundation: a review. Geomorphology, *90*(3), 208-225.

[49] Nkwunonwo, U. C., Whitworth, M., & Baily, B. (2019). Urban flood modelling combining cellular automata framework with semi-implicit finite difference numerical formulation. Journal of African Earth Sciences, *150*, 272-281.

[50] Adebayo, W. A. (2014). Environmental law and flood disaster in Nigeria: the imperative of legal control. International Journal of Education and Research, 2(7), 447-468.

[51] Cummings, S., Regeer, B., de Haan, L., Zweekhorst, M., & Bunders, J. (2018). Critical discourse analysis of perspectives on knowledge and the knowledge society within the Sustainable Development Goals. Development Policy Review, *36*(6), 727-742.

[52] Griggs, D., Stafford-Smith, M., Gaffney, O., Rockström, J., Öhman, M. C., Shyamsundar, P., ... & Noble, I. (2013). Policy: Sustainable development goals for people and planet. Nature, 495(7441), 305-307.

[53] Sarvajayakesavalu, S. (2015). Addressing challenges of developing countries in implementing five priorities for sustainable development goals. Ecosystem Health and Sustainability, *1*(7), 1-4.

[54] Xue, L., Weng, L., & Yu, H. (2018). Addressing policy challenges in implementing Sustainable Development Goals through an adaptive governance approach: A view from transitional China. Sustainable Development, *26*(2), 150-158.

[55] Oduwaye, L. (2009). Challenges of sustainable physical planning and development in metropolitan Lagos. Journal of Sustainable Development, *2(1),* 159-171.

[56] Kovacs, Y., Doussin, N., Gaussens, M., Pacoud, C. L., & Afd, O. G. (2017). *Flood risk and cities in developing countries*. Technical Reports Technical Reports, 35.

[57] Hammond, M., Chen, A. S., Batica, J., Butler, D., Djordjević, S., Gourbesville, P., Manojlovic, N., Mark, O., & Veerbeek, W. (2018). A new flood risk assessment framework for evaluating the effectiveness of policies to improve urban flood resilience. Urban Water Journal, *15*(5), 427-436.

[58] See, L. S., Calo, L., Bannon, B., & Opdyke, A. (2020). An Open Data Approach to Mapping Urban Drainage Infrastructure in Developing Communities. Water, *12*(7), 1880.

[59] Crichton, D. (1999). The risk triangle. Natural disaster management, *102*(3), 1-5.

[60] Kourgialas, N. N., & Karatzas, G. P. (2011). Flood management and a GIS modelling method to assess flood-hazard areas—a case study. Hydrological Sciences Journal–Journal des Sciences Hydrologiques, *56*(2), 212-225.

[61] Liu, J., Wang, S. Y., & Li, D. M. (2014). The analysis of the impact of land-use changes on flood exposure of Wuhan in Yangtze River Basin, China. Water Resources Management, *28*(9), 2507-2522.

[62] Salami, R. O., Von Meding, J. K., & Giggins, H. (2017). Urban settlements' vulnerability to flood risks in African cities: A conceptual framework. Jàmbá: Journal of Disaster Risk Studies, *9*(1), 1-9.

[63] Müller, A., Reiter, J., & Weiland, U. (2011). Assessment of urban vulnerability towards floods using an indicator-based approach--a case study for Santiago de Chile. Natural Hazards & Earth System Sciences, *11*(8). 2107-2011. doi:10.5194/nhess-11-2107-2011

[64] Tapia, C., Abajo, B., Feliu, E., Mendizabal, M., Martinez, J. A., Fernández, J. G., Laburu, T., & Lejarazu, A. (2017). Profiling urban vulnerabilities to climate change: An indicator-based vulnerability assessment for European cities. Ecological Indicators, *78*, 142-155.

[65] Cutter, S.L., Boruff, B.J., & Shirley, W. (2003). Social vulnerability to emvironmental hazards. Social Science Quarterly, 84(2), 242-261.

[66] EC (European Commission). (2004). *Flood risk management - Flood prevention, protection and mitigation.* Communication from the Commission to the Council, the European Parliament, the European Economic and Social Committee and the Committee of the Regions.

[67] De Moel, H., van Alphen, J., & Aerts, J.C. (2009). Flood maps in Europe – methods, availability and use. Natural Hazards and Earth System Sciences, *9*. 289-301

Chapter 11

The Cultural Differences in the Perception and Application of Crisis Management in Tourism

Marica Mazurek

Abstract

During the COVID 2019 outbreak countries in the world reacted to the epidemic situation differently. These discrepancies were based on the cultural differences and the reactions of public sector to deal organizationally and financially with these negative externalities, which can damage also tourism businesses. In this book chapter has been explained the differences in the reactions of Eastern cultures and Western cultures and their hierarchical approach to the decision-making process. The methodological approach to this book chapter and its content is based on the use of concepts rooted in the studies of applied models of crisis management and the application of several case studies from Europe, Asia and North America, where has been discussed the preparedness of public sector to bear a risk and to act effectively during COVID-19 outbreak. A discussion comprises cultural differences and their impact on health situation and the role of media as well as the organizational learning culture.

Keywords: cultural differences, crisis management, organizational learning, pandemic situation of COVID-19, tourism

1. Introduction

The existence of danger and threat are serious factors, which could undermine image, reputation, competiveness of tourism and the whole country. In the global crisis situation, a majority of countries has to face risk of health and human´s security, but also the economy survival. Responses of different countries to these events depend on a variety of factors, especially the economic position, the model of governance, preparedness to cope a critical situation, reaction of international community, mass media, and business culture.

During the COVID 2019 outbreak, communities seem not to be fully and similarly empowered and organizationally and financially prepared to cope with these negative externalities, which can damage also tourism businesses. Eastern cultures do not react the same way as western cultures and their hierarchical approach to the decision-making process could be a strong argument that generic models or approaches would not be implemented in the same way in different cultural milieu, which has also a strong influence on the organizational learning. Similarly, there could be differences even in the reaction of countries joined in common geographical and political structure. For this reason, it might be interesting to study some discrepancies in the reactions of those countries and their managerial preparedness and the organizational specifics for a critical situation especially in tourism due to the pandemic outbreak of COVID 19.

Safety and security are important factors of competitive advantage of a destination, which might not only serve as a place of existence and life of humans, fauna and flora, but also a place for economic and social activities, which are typical for tourism. Those factors are not eternal and unchanging, which is a real danger for the competitiveness, but also the existence of these above mentioned subjects or elements. One of the most vulnerable activities, which might be influenced by safety and security hazards, is tourism. The most important is to understand different patterns of the same problem, which was created by a crisis, and to distinguish the difference of the approaches of different cultures and countries to the same problem and learn a lesson of the organizational differences based on a variety of cultural approaches. Tourism destinations are as vulnerable as any other places, and sometimes more so, and for this reason the crisis management will discuss specifics and organizational learning tasks also from this point of view.

The methodological approach to this book chapter and its content was framed by the conceptual base of studies of applied models of crisis management and the responds of several studied countries to the pandemic situation of COVID 19, especially the preparedness of public sector to bear a risk and to act effectively. A discussion comprises cultural differences and their impact on health situation and the role of media as well as the organizational learning culture. Organisational learning was found to be a critical source of sustainable competitive advantage [1] as stated by Škerlavaj et al. [2]. There will be discussed the questions of tourism in the connection to the economic consequences of pandemic situation. The case studies will be based on the studied secondary sources in selected countries in Europe, Asia, and North America.

2. Conceptual base

The concepts dealing with the crisis management portfolio deal with the reasons or the impacts of crises and disasters. An important perspective to study and understand is the perception of the crises and their solution, which means the preparedness to cope a disaster, set priorities and responses of countries and communities to crises. Faulkner ([3], p. 139) mentioned that "different internal cultures and modus operandi become barriers to communication and co-operation between organizations". It concerns countries, their governments, people, businesses, social groups and tourism as one of the business and social activity as well. For this reason, it is also complicated to apply the universal model for the crisis management.

Hofstede [4] mentioned in his work that people from different national cultures tend to have different styles of management. Based on the author ([4], p. 28) "in the process of comparing phenomena similarity and differences are two sides of the same coin; one presupposes the other." It concerns not only people, but also the institutions position, role, involvement. Important work from Hofstede [4] is the idea to take into consideration the division of societies in the world into the individualistic or collectivistic cultures, which has an impact on people's behavior and the approach of the whole society and government to the urgent tasks in society. Hofstede [4] explained five dimensions of national culture, which influence a behavior of different cultures and it means also countries with people living predominantly in this cultural group. Those typical independent dimensions are: power distance; uncertainty avoidance; individualism versus collectivism; masculinity versus femininity; and long-term versus short-term orientation. Škerlavaj et al [2] mentioned that only a few studies have applied Hofstede´s model to examine the effects of national cultural dimensions on organizational learning.

Among the above mentioned dimensions, power distance means the hierarchy of power and wealth among the general population and a nation, culture, and business. A higher degree means a higher hierarchy, which is executed in society. It allows governments to imply more easily a power in society. It might influence the role of public sector versus private business and concerns the differences of aims of public and private enterprises and their organization. In the connection to the crisis management execution, the role of public sector is unquestionable; however, the scenario of mutual roles of both sectors depends not only on the power distance factor, but on the type of government's response to the crisis, which could be for instance the influence of tighter centralization in a country. Organizational learning from this situation will be based on the direction in a particular country and the role of private and public sector in crisis management.

The uncertainty avoidance could be defined as the affinity to the status quo, less change in society, tendency to keep strict codes and obey the rules in society. The feeling of absolute truth might be a reason for further dictatorship from the side of government, which might complicate free entrepreneurship provision. Less tolerance in society might influence the behaviour of companies and organizations in a country.

Individualism versus collectivism means a preference of being more independent and less governed or on the other hand better compatibility with the other members of society, families, friends, etc. Uncertainty avoidance means a fear of unknown or not certain situations and it might influence also decision level and empowerment in society. In such situations as health risk it could influence behaviour in a positive or negative way. This type of behaviour influences the speed and type of changes in society, business environment and changes, which should be done really smoothly, quickly, and in a more massive way due to crisis situation.

According to Hofstede [4] as stated in Compiranon and Scott [5], individualism stands for a society in which the ties are loose between individuals, and as a result, individuals are only expected to look after himself/herself and his/her immediate family. Conversely, collectivism stands for a society in which people from birth onwards are integrated into strong, cohesive in-groups, which throughout people's lifetime continue to protect them in exchange for unquestioning loyalty. In management decision making and organisational learning situation, a collective decision is preferable in a collectivism culture, whilst an individual decision is more likely to be seen in a culture that supports individualism. This might complicate even decisions of government in the area of health protection and risk avoidance, which could be generated by such a negative externality as the pandemic situation (as one possible outcome of risk management situation), which has consequently negative influence on the whole country, quality of life, security, economy where tourism business is part of it.

Division of roles between genders is incorporated in the expression of masculinity versus femininity and this could be also applied in crisis management concept and organisational learning and decisions in a country. As Compiranon and Scott [5] explain the ideas of Hofstede [4] masculinity is found in a society, in which social gender roles are clearly distinct; thus men are encouraged to be assertive, tough and focused on material success. Women are expected to be more modest, tender and concerned with the quality of life. Unlike masculinity, femininity stands for a society in which social gender roles overlap, and both men and women are encouraged to be modest, tender and concerned with the quality of life. Hofstede [4] explained how masculinity and femininity approach influences culture and as a consequence how managers in a femininity culture prefer to use more intuition, deal with feelings and seek consensus. In masculinity culture, the managers are more decisive, firm, assertive, aggressive, and competitive. More masculine societies are focused on achievements, material rewards and success, which influences also the learning about the business culture in such countries and underlines a type of behaviour of managers who want to succeed in their business strategies.

The question is how this might influence the crisis management process and organisational learning, consequently also tourism business in those countries having a more masculine or feminine dominance society. In COVID-19 crisis situation, surprisingly the countries with more feminine culture impact (Scandinavian countries for instance) achieved better results in fighting the epidemic situation. It might be a result of preferring health protection over business, at least in the beginning of the crisis situation.

The authors Compiranon and Scott [5] discussed the role of culture and leadership and described the crisis management stages in the following scheme (**Figure 1**). They used the ideas of the World Tourism Organization Model. The following scheme shows the main ideas.

Eastern cultures do not react the same way as western cultures and their hierarchical approach to the decision-making process could be a strong argument that generic model would not be implemented in the same way as it would be in western societies.

Some form of criticism also lies in adoption of similar management methods and organizational decisions to different management environments. "For example, the authors FanN and Zigang [7] compared the differences between reaction of American and Chinese managers while dealing with uncertain situation: "having a high uncertainty avoidance culture, Chinese managers normally lack and adventurous spirit and the sense of risk. On the other hand, low uncertainty avoidance American managers are more likely to accept risk." These examples only confirm what the other authors discussed as being in impertinent situation for implementation of models in different environments. Thus, academics as Faulkner mentioned this possibility by stating that "different internal cultures and modus operandi are barriers to communication and co-operation between organizations" [3]. In a case of the epidemic situation; however, we have to face a totally different situation and

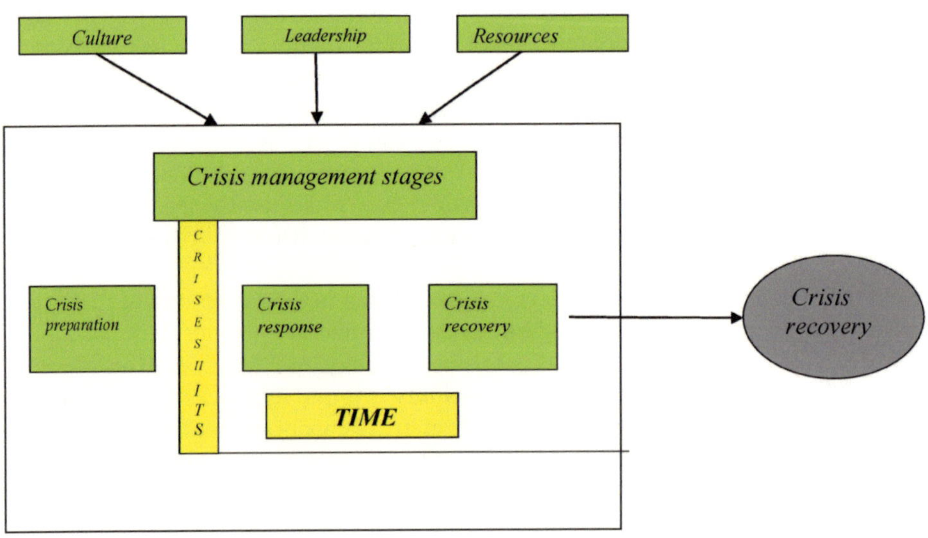

Figure 1.
Crisis management stages. Source: amended upon Compiranon and Scott [5] and the World Tourism Organization [6] model.

it is quite smart to ask if we should be adventurous or more predictive and cautious in order to save somebody's health and life. There is always an open question if the health is a priority or the economy, business, for instance also tourism business. Many countries were able to make reasonable decisions to save both or just to do their best for citizens, their health, but also the existence of businesses and survival of the economy and tourism as well.

Compiranon and Scott [5] agreed "that national culture has a significant impact on crisis management." Johnson and Peppas [8] stated that "crisis intensity varies from country to country and culture to culture, which means that it is very important that crisis response plans are developed for a specific location." It influences a society as social and economic structure with such an economic phenomenon as tourism, a role of government in a society, a role of people as social entities and their culture and behavior, and a role of media as a mean of communication in a society.

The authors Faulkner [3], Ritchie [9], Paraskevas and Arendell [10] mentioned the role of mass media during the crises and disasters. Media role is closely related to image and reputation. The connotation of meaning of crises and disasters can be positive and negative; however, predominantly negative. Though, in Chaos Theory, the existence of a "turning point can be "essentially creative, rather than a destructive process" as described by Faulkner ([3], p. 137). The author explained several examples of this positive outcome as for instance "the empowerment of a society, the creation of modern facilities, innovations, international recognition of destination, etc." It might be really disputable if this could be a case of health pandemic situation in the globalized world, but it should be mentioned also this opinion in order to understand some developments and changes especially influenced by the processes of innovation in the world. As Compiranon and Scott [5] explained the ideas formerly delineated by Holmes [11] that at the heart of every crisis lies tremendous opportunity, and perhaps this is why the Chinese word for crisis is surprisingly composed of two symbols of meaning 'danger' and 'opportunity'. For this reason, it might be important to see and predict which countries might be more in a danger and which will take the opportunities and the same could be visible in the business sphere and tourism could be one example. For instance, tourism businesses, which might be more friendly with modern technologies, digitalization or countries, which are not so tightly depended on mass tourism development and are more typical in a sustainable tourism development, would have probably easier way for the adjustment to a new situation and a real change of business strategies.

Culture, resources and leadership (political and economic), geographical character (for instance isolation as more the islands can use as their advantage in this concept), time (which is now visible in the development of the pandemic crisis, stages and waves of the crisis), level of preparedness, responses of governments, citizens, businesses, especially power of economy, it all might have an enormous influence on crisis recovery, and for this reason could be visible also differences in several parts of the world and also in tourism business performance and changing preferences and visitors' behaviour.

In order to understand the questions of crisis management generally, but also in tourism, some authors tried to develop generic model suitable not only for tourism destinations, but also for different purposes, for instance a country generic model with its specific requirements of safety, security, service provision. Faulkner [3] applied a deep inductive approach in order to construct generic model of crisis and disasters.

Ritchie [9] underlined a necessity of more holistic and strategic approach. Hence, models are more useful for studies of the first group of conceptual approach, e. g. the reasons of crises and disasters and the roles of stakeholders during these events. However, some authors as Paraskevas and Arrendell [10] shifted further risk assessment research of crisis management to the different methodological approach by questioning particular destination stakeholders, corporate and government representatives, policy makers and planners about their preparedness to deal with crises and disasters, which could be an excellent lesson of different approaches to the organizational learning and managerial decisions understanding. As the authors stated, "the purpose was to produce insight rather to test theory, the study was inductive in nature and used a qualitative, interpretative approach" [12]. Their research revealed through interviewing of experts on corporate and government security, safety, tourism policy and planning some controversial aspects of former research approaches based on compiling of theoretical frameworks without testing the attitudes of stakeholders. A research underlined necessity of co-operative approach of all stakeholders, compatible jurisdiction, allocation of financial resources, etc. Thus, pragmatic approach to the studied topic revealed important gaps between managerial theoretical approach, organization, and practice.

More discussion is needed on perception of disaster management framework of models (re-active models) and pro-active risk management models as has been stated by some academics in numerous academic journals dealing with the topic of crisis management. Important role in the crisis management and resolving the situation has a state and its role is crucial. It is well known in the public economy theory that public sector has to be present where the private sector is not capable of solving a problem, but has to withdraw when it is not necessary to intervene. Crisis management is a really difficult role, which should be planned and prepared thoroughly ahead and kept strongly during the occurrence of the negative situation in a country and the world. Many countries and their businesses failed due to unpreparedness and due to risking of health of their inhabitants and the consequences in those countries could be tremendous. For this reason, a discussion about the preparedness and models of crisis and disasters is needed.

First academic, who identified these two approaches to model creation in crisis and disasters, was Heath [13, 14] who mentioned the traditional crisis management approach and the risk management approach. Miller and Ritchie [15] added that "the traditional crisis management approach involves no initial (pre-crisis) planning or management (**Figure 2**) and the role of risk management approach "is to respond to the crisis and manage the impacts effectively and efficiently (**Figure 3**).

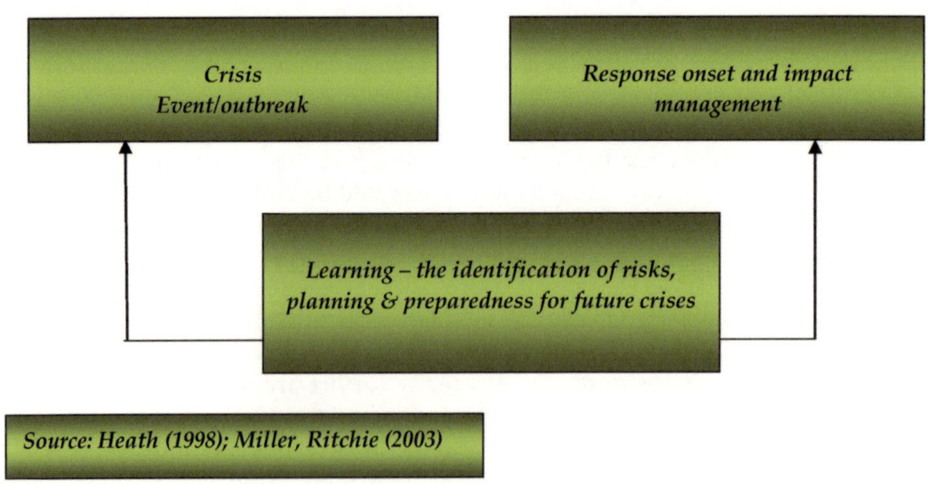

Figure 2.
A traditional approach to a crisis.

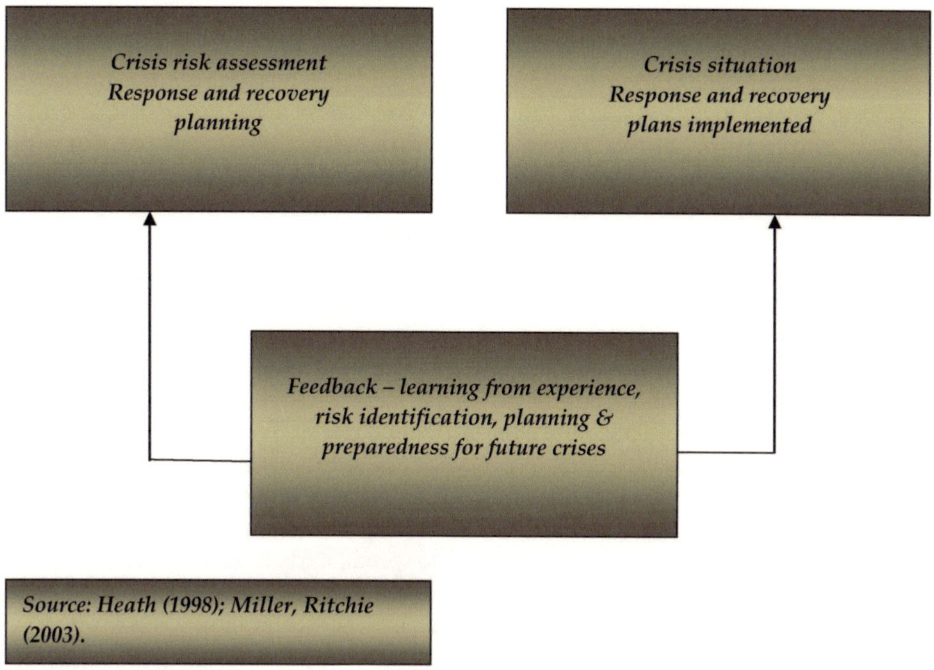

Figure 3.
A risk management approach to a crisis.

The methodological approach is based on the qualitative approach and is framed by the conceptual base of studies of applied models of crisis management and the responds of several studied countries to the pandemic situation of COVID 19, especially the preparedness of public sector to bear a risk and to act effectively and the responds of governments and citizens to the crisis situation. Škerlavaj et al. [2] mentioned that the type of predominant culture would bring diverse influence on the development of organizational learning culture. Crucial are especially cultural differences and a role of media in several discussed countries. The case studies are based on the studied secondary sources in Europe, especially in Slovakia, Czech Republic in comparison to the other countries in Asia (Taiwan, South Korea) and North America (Canada and the U.S.A.), etc.

3. Results

Several studies from Asian countries showed that in many cases could be visible former experience with crisis situation and it means also preparedness of a responsible government to that situation. Moreover, there might be visible cultural dimensions, which have been mentioned as the collectiveness or the individualism. Important could be fast political decisions and a respond of citizens. For instance, one excellent example is Taiwan. Taiwanese government is one of the most successful examples of crisis management implementation in the world. The first information about the virus appeared on 21st of January 2020. Taiwanese government has actively and really efficiently sent all instructions about the protection against a new form of virus to the citizens and did not try to hide any information, which is a sign of democratic and responsible government. One of the crucial tasks was a control of the healthcare supply chain affordable to the country and its citizens and a tight co-operation with the academic institutions in a matter of the antiviral drugs development. For instance, the figures by April 9th, about 79 days after first case appeared, the number of cases was 379 and deaths only 5. These numbers were much lower than the numbers in China in that time, which is a result of a quick response, geographical advantage (an Island separated by a sea), preparedness and different cultural and political approach despite of being Chinese culture, but with a totally different political attitude. When we compare the numbers of the evidence of this virus to the situation in China, by April 9, in China the number of confirmed cases was 175,74 times the number in Taiwan and the number of deaths 5,300 times the number in Taiwan (in Taiwan 4,7 cases and 0,06 deaths per day).

Another example of success might be found in Malaysia and Vietnam. Similarly, as in Taiwan, Malaysia and Vietnam are culturally close to Confucianism. It means that governmental leadership might be easier because of that collectivist feelings and a meaning of collective good is deeper incorporated in their cultures. It means hegemony of duty to society over individual needs. This was visible in those countries, where for instance citizens of Taiwan regularly wear facemasks in public despite of the fact that the evidence of COVID-19 is very low. Governments in Asia need not always remind people to wear masks, keep distance and stay home.

In Malaysia, COVID-19 infection started to spread early March and rocketed to 8 800 cases early June, but later due to the discipline and facemasks, responsibility of the citizens and government regulations obedience, the number of cases dropped. Ethnic Malay cultures in Malaysia and Indonesia promote banding together against common threats.

Malaysia is also one success example of the cultural influence, governmental approach and responsibility of citizens; however, there could also play important role the geographical indicators and a distance from the neighbouring countries. Boundaries, geographical distance, social distance and political capability might be decisive factors of successful outcome of such pandemic situation caused by a virus.

Similarly, Vietnam was able to keep the situation of their country with just 401 cases in the beginning under control; however, there might be visible not only cultural, but also political influence and more governmental control as a consequence of former historical and political development. Despite of it, Vietnam could be a success story to the world.

In the United States, the virus started to develop in early March, but in comparison the above mentioned countries, the numbers have climbed in June 2020. It might be a cultural attitude and power distance characteristics, but Americans are not unified in the rule to wear masks and abandon their personal freedom to decide personally. This might be a problem in several western countries all over the world, for instance also in Europe or even in Eastern Europe (a case of Czech Republic).

On the opposite to the U.S.A., another country at the North American continent, Canada knows as a multicultural country focused on social, health, and community principles. Canada has a different story as for instance the U.S.A. and the rest of western world (particular countries). Based on a research of Zhang and Young-Leslie who have been collecting data mainly through focus groups and surveys of Canadians from across the country, several results could indicate a cultural and political approach as well the attitudes of different cultures to the rules given by Canadian government as a result to the pandemic situation. The research showed that some rules as face masks wearing was quite common even before the COVID-19 outbreak among some cultures, for instance as a result of fair pollution or sensitivity to toxins in the air. Those cultures were from East Asia, for instance China, Korea, Vietnam, and Japan. In some cases, these inhabitants tried to avoid harsh weather or wanted to keep anonymity. In a poll conducted by Leger and the Association for Canadian Studies, 51 per cent of Canadians surveyed said they've worn masks while doing their grocery shopping. Fifty-three per cent said masks should be mandatory in public and confined spaces like shopping malls and public transit. Public acceptance of protective face masks has evolved dramatically in Canada since the beginning of the COVID-19 pandemic, according to new psychology research from the University of Alberta. It is important to state that North-American people, a generation of people who are still alive and the middle aged or young generation never experienced Spanish flu. This could be also stated about a majority of people in the world because this flu was typical for the beginning of the 20th century and not many people are still alive from that period of time. This might be also important fact in general judgment of the behaviour of some people who do not believe in this real health problem and think this might be only made up artificially and distributed by media. However, there are again political, cultural and geographical differences among countries and people.

Zhang and Young-Leslie also found there were differences between non-Chinese-speaking Asian-Canadians and recent immigrants, where the assimilated Chinese non-Chinese speaking Asian Canadians felt to be more targeted as new immigrants, which is also an important sign of a stigma. However, based on this research and results, it could be visible that cultural influence and a period of life in different country and culture might have an influence on behaviour of people.

Richard Schultz, an expert on federalism and a 40-year veteran of teaching politics at McGill University mentioned important statement on Canada, which should be discussed in order to understand the differences among the development in the epidemic outbreak in Canada and the U.S.A. There is this culture (in Canada) of ... more deeply rooted community and social services. We fight about the size of government, we fight about deficits -- but when push came to shove, we said, 'Look, there's no one fighting this.' ... it does say something to me about the vast cultural difference between the United States and Canada. "Professor Schulz continues" Political scholars have long seen Canada as one of the world's most decentralized federations -- a place where Ottawa yields much to the provinces and territories, which manage key services like health and education." (https://www.ctvnews.ca/health/coronavirus/compared-to-u-s-canada-s-covid-19-response-a-case-study-in-political-civility-1.4895357).

However, in the question of crisis solving, there has been a strong consensus and co-operation among the provinces and the federal government. Important is also a consensus with the communities and citizens, businesses and economic support in the time of crisis to overcome the negative consequences. Professor Schulz commented that "And yet, I think this is a highly exceptional case that we're dealing with. We have the 10 provinces and the federal government -- in a way that I haven't witnessed in the 56 years I've been studying it -- working relatively collaboratively, co-operatively together on this issue.", which confirms the above stated ideas (https://www.ctvnews.ca/health/coronavirus/compared-to-u-s-canada-s-covid-19-response-a-case-study-in-political-civility-1.4895357) [16]. It might be more explained by one fact that Canada has had already an experience with SARS outbreak in 2004 and a positive outcome of this situation was preparedness for the epidemiological and crisis situation. Important role might play also cultural factors as has been mentioned above and the fact that Canada is a country with strongly developed common sense feeling.

In Canada, the outbreak of SARS (Severe Acute Respiratory Syndrome) did not have an extreme impact on mortality of people because only 45 people died, but an immediate effect was evident in tourism industry. Over 1/3 of 95 000 employees in tourism was laid off (based on Smith Travel Research) after the SARS outbreak and total decrease of tourism revenue due to SARS was 500 million in Toronto, Ontario in the following months. From April to June 2004, the number of international visitors declined 14%, their spending declined 13% and the travel deficit in the income from international tourism was over 1.1 billion CAD together with the decrease of employment in tourism by 2.4% ([17]; KPMG; PKF Consulting). Based on the Canadian Charter of Rights and Freedoms, Federal government has a power to act in a matter of health protection in a case of health protection of the whole country despite a fact that health care, public health lies under the jurisdiction of the provinces. Some formerly experienced problems and failings during SARS outbreak in 2004 lead to a stronger federalism in this question, which had an influence on Canadian story in pandemic fight. It might be a real problem in the second largest country in the world, but the outcome was not catastrophic and when we compare the situation in the U.S.A., Canada was able to cope the crisis situation much more efficiently. Fierlbeck commented that Canada, because of historical circumstances, really has what I would call a reasonable institutional framework for co-ordination between jurisdictions". (https://www.ctvnews.ca/health/coronavirus/compared-to-u-s-canada-s-covid-19-response-a-case-study-in-political-civility-1.4895357).

The success lesson could be taught from Slovakia in the 1st wave of COVID-19 situation, where mostly several key factors played the most important role, the quick introduction of protective rules, which were especially rooted in wearing protective masks and gloves. The strict rules were implemented in order to protect citizens as for instance a penalty of breaking a quarantine order. Slovakia belonged to the first countries in the world (second after the Czech Republic in Europe) to order face masks to become mandatory inside buildings (stores, schools, etc.) and in public spaces. This decision was made even earlier as the World Health Organization advised people to wear masks in public. By March 13, one week after Slovakia confirmed its first coronavirus case the Slovak governmental representatives appeared in masks in front of media and demonstrated their compassion with the existing situation and the seriousness of the health care problem caused by the COVID-19 virus. The message was sent to the public: “Protect others and you'll be protected ... It's not embarrassing. It helps everyone.” Important decision was a nationwide lockdown. The reason might be a fear of the situation in the world, especially in Italy and Spain and a fear to cope a pandemic situation, which could be overwhelming and devastating for the Slovak healthcare system. (https://www.theatlantic.com/international/archive/2020/05/slovakia-mask-coronavirus-pandemic-success/611545/) [18].

When analysing the success factors of Slovakia in a survey about the successful measures fighting against the virus of COVID-19, the most important were classified the rule of wearing face masks, gloves, especially in very frequent spaces. About 90 percent of the respondents have limited their travelling, either by public transport or by car. This had a strong consequence on tourism and travel agencies and airports experienced a strong decline of passengers (about 80%). (https://newsnow.tasr.sk/featured/survey-over-90-of-slovaks-view-coronavirus-related-measures-as-appropriate/)

Unfortunately, this is not a case of the 2nd COVID-19 wave in Slovakia, where the situation is becoming more difficult. Slovakia and Czech Republic are culturally very close countries in some aspects and at the beginning of the pandemic situation in Slovakia was second after Czech Republic to implement face masks duty after the outbreak of COVID-19 in their countries. Both countries have a democratic government, which was elected in free elections and the development in fighting the epidemic situation was at the beginning similar, despite a slightly higher numbers in Czech Republic due to the number of citizens and a proximity to western countries, which were more affected in that period of time. In the first wave of pandemic situation both countries were cases of good results. In June the situation has been improved and both countries opened the economy, schools and some travelling to safe countries was fully introduced. However, it is visible from the development in both countries that Slovak citizens were more careful in opening and did not abandon some formerly introduced regulations. Slovaks are people who obey the rules and it is more collectivist society with a masculine characteristic. This cannot be fully generalized, but when we compare Czechs and Slovaks, there are differences.

This might be a reason why there exist now such differences in the number of infected people, mortality and 14-day cumulative number of cases per 100 000 when we compare both countries now. In Czech Republic (now takes 2nd place in Europe in the daily increase of numbers of infected people), there are 49 290 cases, daily increase ranks from 2000 to over 3000 infected people, mortality is 503 and 14th day cumulative number of COVID-19 cases per 100 000 is 37,9. The expectation based on the European Centre for the Prevention and Control of Diseases the expected daily increase in Czech Republic could be 8000 cases a day. Finally, the government decided to renew the meetings of the General Crises Committee and decided about a personal change of a Minister of Health Care. In Slovakia, on the other hand are 6 677 infected, mortality is 40 and the 14-day cumulative number of COVID-19 is 37,9 per 100 000. Slovakia had to restrict the travelling rules from Czech Republic and there are several strict restrictions, which will try to avoid spreading the virus. Slovak government tries despite very friendly contacts with Czech government to look at the case as the negative externality, which might be a danger for Slovak citizens. Culturally, Czech people could be characterized as more feminine society (in comparison to Slovakia as more masculine society), more individualistic society closer to western European countries and a society with not such a tendency to obey rules (refusal to wear masks inside, for instance) and keep all restrictions, especially in big cities. Cultural dimension, political rule, governance and also the number of visitors with tourism or business aim might be a decisive reason for Czech Republic to be in such a situation. In all aspects, economy and consequently even tourism suffers more when people are not administered properly or there is lack of control from a government. Obviously, this pandemic situation might lead to stronger governmental role in a country and in tourism business as well. It might be a lesson for the countries and governments of those countries how to solve the situation more effectively. The effective crisis management and organizational learning processes should be helpful not only to understand the differences among cultures, but especially could solve problems in a faster and progressive way.

4. Conclusion

Competitiveness of countries, which is based not only on comparative advantage, but also the competitive forces as for instance is safety and security, has tremendous impact on economy and tourism as well. The world is in continual change, which could be positive or negative. Some changes might be totally unexpected and devastating for the economy and the most dangerous are consequences for the human´s health and life, which is a case of pandemic COVID-19, which affected the whole world since January 2000. In this chapter, we tried to discuss not only managerial preparation and the existence of models of crisis and management from former crisis situations, but also preparedness of several countries to cope critical situation, the role of mass media and business culture and especially the influence of cultural differences in managerial decisions, in behavior of citizens generally and in the discussed countries. Hofstede [4] explained five dimensions of national culture, which influence a behavior of different cultures and it means also countries with people living predominantly from this cultural group. Those typical independent dimensions are: power distance; uncertainty avoidance; individualism versus collectivism; masculinity versus femininity; and long-term versus short-term orientation. Škerlavaj et al [2] mentioned that only a few studies have applied Hofstede´s model to examine the effects of national cultural dimensions on organizational learning. For this reason, we tried to discuss if those mentioned dimensions could have an influence on the development in crisis situation in the studied countries in the 1st wave of COVID-19 (not including the 2nd wave or the period after 2nd wave with new mutations of the virus COVID-19). It is evident that for instance high power distance culture would enhance the positive effects of information interpretation, information acquisition and behavioural and cognitive changes as the important variables of organizational learning, but on the other hand the individualistic, masculine and the uncertainty avoidant culture would weaken or hinder such process. For instance, in such situation as crisis, lack of flexibility caused by the uncertainty avoidant culture could be dangerous for crisis problems solutions as well as for the organizations who are not able to learn from a failure, do not engage experimental learning and would hinder the development of the organizational learning culture. These several examples could be important for the statement that the roles of national culture could be decisive for the organizational learning culture and that different cultural dimensions influence organizational learning culture. Dimensions of national culture could have an impact on the whole process of crisis management. For this reason, the same situation cannot be totally the same in every country despite of taking similar restrictions or providing similar processes of crisis management and organizational learning. Consequently, the situation in risk environment has an impact on economy (unemployment, bankruptcies of businesses, social problems, etc.). Tourism is a part of social and business environment by its activities and goals and a destabilizing situation in the world has a really negative consequence not only on humans, countries, but also tourism businesses.

Author details

Marica Mazurek
University of Žilina, Slovakia

*Address all correspondence to: marica.mazurekova@fhv.uniza.sk

IntechOpen

© 2021 The Author(s). Licensee IntechOpen. This chapter is distributed under the terms of the Creative Commons Attribution License (http://creativecommons.org/licenses/by/3.0), which permits unrestricted use, distribution, and reproduction in any medium, provided the original work is properly cited. (cc) BY

References

[1] de Gues, A. P. (1988). Planning as learning. Harvard Business Review, 88, March-April, pp. 70-74. In Škerlavaj, M, Huang, M. & Su, C.(2013) The moderating effect on national culture on the development of organizational learning culture: A multilevel study across seven countries. *Journal for East European Management Studies*, pp. 97-134.

[2] Škerlavaj, M, Huang, M. & Su, C.(2013) The moderating effect on national culture on the development of organizational learning culture: A multilevel study across seven countries. *Journal for East European Management Studies*, pp. 97-134.

[3] Faulkner, B. (2001). Towards a framework for tourism disaster management. *Tourism management,* 22 (2001), 135-147.

[4] Hofstede, G. (2001). *Culture's Consequences*, 2nd edn. Thousand Oaks, California: Sage publication.

[5] Compiranon, K. & Scott, N. (2007). Factors influencing crisis management in tourism destinations. *Crisis Management in Tourism*. 2007, 142 – 156.

[6] World Tourism Organization (2005a) In Compiranon, K. & Scott, N. (2007). *Factors influencing crisis management in tourism destinations. Crisis Management in Tourism.* 2007, 142 – 156. Crisis Guidelines for the Tourism Industry, World Tourism Organization, (http://www.worldtourism.org/tsunami/eng.html).

[7] Fann, P. & Zigang, Z. (2004). In Campiranon, K. & Scott, N. (2007). Eds. Factors influencing crisis management in tourism destinations. *Crisis Management in Tourism*, 2007, pp. 142-156.

[8] Johnson, V., Peppas, S. (2003) Crisis management in Belgium: the case of Coca-Cola. *Corporate Communications* 8, 18-23. In Campiranon, K. & Scott, N. (2007) eds. Factors influencing crisis.

[9] Ritchie, B. W. (2004). Chaos, crises and disasters: a strategic approach to crisis management in the tourism industry. *Tourism Management*, Vol. 25, pp. 669-683.

[10] Paraskevas, A. & Arendell, B. (2007). A strategic framework for terrorism prevention and mitigation in tourism destinations. *Tourism management*, 28 (2007), 1560-1573.

[11] Holmes, J. (2003, May 2005). *Asia Pacific Business Opportunities, International Congress and Convention Association*. http://www.iccaworld.com.

[12] Saunders, M., Lewis, P. & Thornhill, A. (2003). Research methods for business students (3rd ed.) Harlow: Prentice-Hall. In: Paraskevas, A. & Arendell, B. (2007) eds. A strategic framework for terrorism prevention and mitigation in tourism destinations. *Tourism management, 28* (2007), p. 1560-1573.

[13] Heath, R. (1998). The Kobe earthquake: Some realities of strategic management of crises and disasters. *Disaster prevention and management*, 4(5), 11-24.

[14] Heath, R. (1995). The Kobe earthsquake: Some realities of strategic management of crises and disasters. *Disaster prevention and management*, 4(5), 11-24.

[15] Miller, G., A. & Ritchie, B., W. (2003). A farming crisis or a Tourism Disaster? An analysis of the Foot and Mouth Disease in the UK. *Current Issues in Tourism*, Vol. 6, No 2, 150 – 171.

[16] McCarten.J. (2020, September 3). *Compared to U.S., Canada's COVID-19*

response a case study in political civility. https://www.ctvnews.ca/health/coronavirus/compared-to-u-s-canada-s-covid-19-response-a -case-study-in-political-civility-1.4895357

[17] Wall, G. (2006). *Recovering from SARS: The Case of Toronto Tourism*. In: Tourism, Security and Safety. Oxford: Elseviere.

[18] Serhan, Y. (2020, September 3). *Lessons From Slovakia—Where Leaders Wear Masks*. https://www.theatlantic.com/international/archive/2020/05/slovakia-mask-coronavirus-pandemic-success/611545/)

Chapter 12

Italian Crisis Management in 2020

Luisa Franchina, Alessandro Calabrese, Enrico Scatto and Giulia Inzerilli

Abstract

Approaches to risk analysis, crisis management and resilience enhancement for Critical Infrastructure (CI) Protection will be considered starting from a case study related to the management of the pandemic in Italy. Business continuity and crisis management models for CI are analyzed aiming to deal with complexity and reduce uncertainty relating pandemic and long-time crisis. Furthermore, is presented a methodology highlighting the functioning of the Italian Civil Protection and its systemic nature: a complex apparatus made up of different elements and organizations, which derives from the functioning of different organizational systems in interaction with each other. As a baseline for the coordination management the Augustus Method is considered for its strategical, tactical and operational aspects. One of the main outputs of the research consists in creating a "what if" forecasting model, configured as a visualization of the propagation of negative effects on the supply chain and manpower over time.

Keywords: complex system, emergency management, manpower, cascading effects, resilience

1. Introduction

Dealing with complexity and reducing uncertainty during 2020 crisis is a priority, for Countries, Critical Infrastructures, and companies.

Due to the interdependency of Critical Infrastructures, companies, and the civil society their protection and management represent a significant challenge and, somehow, an opportunity.

The present contribution aims to support the understanding of the tangled pandemic scenario, studying the interdependencies between different sectors and their supply chain, proposing a model addressed to the complexity management for ensure the Business Continuity both of Critical Infrastructure and companies.

The Italian response to the crisis generated by the pandemic was observed, from the study of the impact of the crisis on Critical Infrastructures, to the response strategies, the remediation plans, passing through the reference standards on business continuity and supply chain (in the ISO family of standards).

The imposed lockdown has led to a forced acceleration of digitization, with the challenges and opportunities that could be derived from it.

The crisis management, supported by the experience generated by the avian influenza, together with the support tools provided by the Italian government has proved to be effective and efficient, also relaunching several SMEs through their productive conversion.

The human factor has become evident as the cornerstone of any service, from the provision of essential services falling within the competence of the Critical Infrastructures, which have involved a particular attention to the continuous security and business protocols to be followed, to the most disparate production sectors. It is also necessary to remember how the interconnection between the different sectors and services now characterizes our reality, and therefore how the so call "What-If Analysis" s fundamental in the development of decision support tools for crisis management. In this context is clear that resilience is founded on risk analysis and the drawing of recovery plans, together with measures for an increased control over the value chain.

2. Addressing complexity and impacts of pandemic in critical infrastructure

Dealing with complexity and reducing uncertainty during 2020 crisis is a priority for Countries, Critical Infrastructures, and companies.

Complexity could represent a risk but also an opportunity to create a new competitive advantage.

Society is dependent on composed critical networks, becoming more complex as are strong interdependent both within and between infrastructure systems [1].

Nowadays, complexity and uncertainty assess the search for new and effective management strategies and methods. Embracing unpredictability and planning to adapt is crucial to manage the complexity that cannot be eliminated, although, it can be reduced to manageable levels. Complexity and vulnerability of Critical Infrastructure systems has been explored and assessed [2, 3].

Complexity is related with composite systems and problems that are dynamic, unpredictable, and multi-dimensional. It consists of a collection of interconnected relationships and parts. Unlike traditional "cause and effect" or linear thinking, complexity science is characterized by nonlinearity [4]. Complexity management needs to consider several layouts of complexity, in fact an IC or a company internal value chain is strongly dependent on external complexity.

For each area of complexity regulation, as avoidance and reduction related to causes, transfer, and division, exist several theories, approaches and methods.

Effective complexity management aim to develop an appropriate and effective incident response plan. Finally, complexity must be addressed proactively.

In fact, in such complex scenario, different actors (institutional and non) have responded to the crisis in multiple ways, according to the regulations issued. Moreover, these troubled times show how strategic and essential are some sectors.

In the crisis generated by the pandemic it has been confirmed that the daily life of the citizens depends on the reliability of the Critical Infrastructures (CI) to supply essential services such as energy and water. In recent years, Critical Infrastructure control systems have become more complex, with increasingly interconnected devices; a trend that will probably continue with the Internet of Things.

The need for increased resilience to resist extreme events of both natural and malicious origin has become more acute. With Critical Infrastructure continuously exposed to threats, especially cyber-attacks, there are severe security implications, most notably in the energy sector which is ranked as one of the most affected sectors with the highest incident costs [5]. Any attack of this nature is likely to have knock-on effects on a country's overall economy and the lives of its citizens.

The pandemic, all in all, has had modest effects on the electrical service. Electricity consumption has been reduced by about 10% on average, but with a very uneven distribution on the Italian territory. Fortunately, the phenomenon has been well controlled and there have been no perceptible effects, but it is easy to imagine the consequences of possible inefficiencies. The effect of the pandemic could be very marked on geopolitical balances, in a context of possible tensions deriving from the rebalancing of the primary energy market and the challenge of the Fourth Industrial Revolution (4IR) [6].

The energy issue brings us back to the more general field of critical infrastructures: electricity and energy system, communication networks, infrastructures for the transport of people and goods (air, sea, rail and road), health system, economic-financial circuits, administrative and state organizations and bodies.

What happened on the Istituto Nazionale della Previdenza Sociale (the Italian Social Security), website is a symptom of a strong criticality in the Country System, where technical shortcomings make the fundamental rights of citizens even more vulnerable, and how IC and companies must equip themselves to manage crisis situations that are not predictable. For this reason there have been several episodes in Italy which have triggered the alarm by the Centro Nazionale Anticrimine Informatico per la Protezione delle Infrastrutture Critiche (CNAIPIC - National Anti-Crime Information Centre for the Protection of Critical Infrastructures).

In terms of crisis management, thanks to the experience of avian influenza (H5N1), which has highlighted how the human factor is the most valuable element for any company and as such must be safeguarded and protected, operators of critical infrastructure have been able to develop a series of effective initiatives, as demonstrated by the fact that no essential service, i.e. the supply of gas, water, electricity, transport, etc. has suffered interruptions or dysfunctions in recent months. And this despite the problems related to difficulties in supply, reduced mobility, the presence of staff in quarantine fiduciary and/or infected and considering the commitment of companies to safeguard the health of their workers.

This achievement is the result of an effort which in recent years has seen a significant change in the role of the security managers, which has shifted to the top management in order to bring strategic choices back to specific task forces capable of having a prompt impact on all levels of the company's organization, being equipped with the financial and decision-making capacity appropriate to the criticality of the situation [7].

2020 long time crisis and consequent lock down were managed asking to every operators of critical services to maintain business continuity and to guarantee services if critical. This means that not only critical infrastructures at national level, but also critical infrastructures at regional or city or province level had to maintain operation, even having the supply chains partly or completely locked and also even having manpower partly or completely in smart working.

The Office of the Military Advisor of the Presidency of the Council, in consideration of the necessity to guarantee the essential services provided by Critical Infrastructures, has provided the precautionary principles, to which Critical Infrastructure Operators are required to comply in order to contain and contrast the spread of the pandemic, while ensuring the continuity of the supply of essential services, the operability of the facilities and the security of the personnel involved.

These lines suggest, first, a reduction in the number of staff working in situ by reducing activities to those that cannot be postponed for business continuity, and to review the maintenance programs, limiting them to those that cannot be postponed and postponing those that are not indispensable, promoting the adoption of smart working at all levels, necessary for the continuity of the service. The Precautionary Principles highlight the need to provide specific training and tools to operators to prevent and combat the threat of cybersecurity, the importance of which is growing today, to equipping all staff with adequate IT support, including the use of dedicated connections, VPN systems and anything else in order to ensure adequate levels of cybersecurity, including the issue of appropriate rules of conduct by staff working in smart working mode.

Furthermore, is required to prepare all the necessary measures related to sanitization.

The Companies are invited to organize the personnel involved in activities that cannot be postponed at the work sites or field operations in teams composed of the minimum number of people necessary for the safe execution of the various activities. The composition of each team, to increase its resilience, must not, where possible, change over time and specific procedural measures must be taken to avoid, or limit to a minimum, physical interaction between several teams.

With regard to the management of the control and management rooms, given that it is necessary to ensure their functionality in all conditions, it is recommended that all useful measures be taken to contain the pandemic; organizing the staff into several teams and adopting specific and more stringent safeguards for this type of personnel, for example, measures and/or adequately equipping several rooms, possibly in different locations, to allow the alternation of shifts in different rooms and/or sanitized each shift change [8]. Another taken measure was the "voluntary segregation": the provision of temporary accommodation for groups of people who will operate in the control center for a period of not less than 14 days without physical contact with external personnel. The spaces to which such staff have access will be forbidden to those who do not implement voluntary segregation. To guarantee the continuous rotation of the activities, a second team of staff is set up at the same time, already in isolation at their homes.

Telespazio has set up a three-level system for its Space Center which, before entering the control room, requires a further period of voluntary quarantine within a camp facility located at the Fucino site [9].

The theme of cyber-security is particularly relevant in an increasingly interconnected world where threat vectors multiply and can affect the vulnerabilities of Critical Infrastructures. Moreover, the low level of cybersecurity preparedness of the country system is also reflected in low awareness among citizen-users.

In view of the above, we can say that for the management of emergencies and crises first of all it is necessary to develop a culture of security, supported by the necessary tools and strategies, also considering that we are moving towards the increasing digitalization of any area of the country. In order to do this we can combine the creation of high potential and distributed networks, to avoid in case of stress of infrastructure use, domino effects. It is not possible today to imagine an area of the country that is not covered by essential infrastructures and services that respond to adequate minimum levels of service delivery and security, especially cybersecurity.

It is therefore also essential to start a training process in line with the needs of the world of work and thus adapt to the new professions, together with a plan for the conversion of skills towards new professional qualifications [10].

A fundamental and new aspect of this crisis, which has led to a rethinking of the management of Critical Infrastructures, is that there was a clear " day before" (in Italy between 10th and 11st March) and a lack of clarity in the "day after". There is still the sensation of a prolonged crisis and the passage to a remote working that has reduced social relations. This situation has also led to a discontinuity in the visibility that the employer has towards his employees (with respect to how he is and what he feels) that had never been experienced before, while the knowledge of the human model is crucial.

When we make a reading of complexity, we consider a company (or a CI) and analyze it in all that is the flow of its value chain and we retrace all the places and moments of a not physiological complexity.

One type of challenge for Critical Infrastructure Protection is about the dependencies and interdependencies among different Critical Infrastructures [11].

In the context of this extremely long lock-down we had an enormous complexity of relations with suppliers and with those who had to remain in continuity and we would find in the re-opening a strong discontinuity, also in understanding for example the rules with which it was possible to re-open and the responsibilities (in fact, the provision of a suitable team that knows how to interpret the rules is also part of crisis management).

2.1 The Italian production strategy during the 2020 pandemic: statal measures and production conversion

Italian SMEs have worked out an appropriate response strategy to the crisis caused by the 2020 pandemic.

Starting from the importance of the role of each individual entrepreneur, through the constant and daily collection of information on a formal and informal basis, it was possible to identify the strategic levers and focus on new core businesses, based on corporate liquidity, assets and resources.

It emerged that the creation of balanced strategic levers, the make/buy balance, together with the dialog with the stakeholders represented a fundamental element for the conception of a response strategy that represented an example of business resilience.

The crisis has certainly been, and still is, an opportunity to examine which lessons are learning for the future creation of resilience-oriented protocols [12].

There are many Italian companies that have reacted to the crisis by reconverting their production.

Phase two, co-existence with the pandemic, began on 4th May 2020. The Prime Minister's Decree issued by the Government has made mandatory the use of the mask in closed places accessible to the public, such as public transport and shops. Wearing the mask is mandatory in all situations where "it is not possible to continuously guarantee a safe distance" [13].

Given the emergency and lack of access to this personal protective equipment, more and more companies have chosen to make a concrete contribution and boost their activities after the lockdown by aiming at the reconversion of production chains to manufacture masks. Initiatives that are born to make available the expertise and skills of entire sectors forced by the emergency and the upheaval of daily habits to rebuild their missions and restructure their short, medium- and long-term objectives.

Siare Engineering, an Emilian company specialized in the manufacture of lung ventilators (the unique company in Italy), at the outbreak of the emergency increased its production and changed its export market. In mid-March the company delivered 300 machines to the Civil Protection, originally destined for countries such as South Korea, India, the Philippines and Vietnam, its traditional clients. The company was supported by specialized Army technicians with the aim of producing over 2300 machines, tripling production. Siare Engineering's efforts were supported by companies such as Ferrari, FCA and Magneti Marelli [14].

Grafica Veneta, a Paduan company active in the printing sector, has reconverted its production to produce 2 million masks. These products, even though they could not be intended for healthcare workers, provided (at a time of dramatic shortage) an initial protection to the population, and were distributed free of charge to the population by the Civil Protection and the Alpini (Italian Army's mountain infantry).

Mestel Safety, a specialist in snorkeling and diving masks, deposited a patent at the beginning of March to transform this diving equipment into protective masks against contagion [15].

On 23rd March Confindustria Moda launched an adhesion campaign to make masks and PPE, to which 200 companies have immediately joined. A similar initiative was taken by CNA Federmoda. Some of the most important Italian fashion companies responded to the call, such as Armani, Calzedonia, Fendi, Gucci and Valentino.

Prada, on request of the Tuscany Region, has started the production of 80,000 white coats and 110,000 masks [16].

Toscano Alta Sartoria (ex Mabro) has promptly reconfigured its production starting, from March, to produce 3000–4000 masks per day [17].

A choice made also by Valigeria Roncato, a leading company in the sector in the production of luggage made in Italy, which has decided to make a strong contribution to the enduring battle at pandemic by converting its production lines for the production of long-lasting, non-disposable, washable and therefore reusable masks [18]. The core business of the Veneto industry responds to the urgent demand for protective masks that are becoming more and more indispensable.

These solidarity initiatives have been stimulated by the possibility to access incentives to activate the production and supply of medical devices and personal protective equipment (PPE) for the containment and fight against the epidemiological emergency.

And more: to deal with the pandemic, numerous measures have been taken to prevent and contain its expansion and its effects on the economic system. These are emergency measures issued at short distance from each other and linked to each other.

The financial support to SMEs has gone through interventions on the fiscal side, the suspension of the refund of loans, the public guarantee on those granted to companies that have suffered decreases in turnover, a fund for the promotion of Made in Italy, financing.

The objective was to prevent SMEs from shutting down due to lack of liquidity because of the emergency: according to Cerved the system could lose up to 650 billion in revenue between this year and the next.

In this picture, are extremely important the interventions to support the liquidity of the productive network, strongly strengthened by the Legislative Decree n. 23/2020 (so-called Liquidity Decree). This last measure has on one hand modified and on the other hand implemented the extraordinary measures introduced by Decree Law no. 18/2020. This is also thanks to the new regulatory framework for State aid, the EU Commission's "State Aid Temporary Framework" [19], which has intervened in the meantime. On 14th April 2020, the European Commission authorized the extraordinary support aid schemes provided by Decree Law no. 23/2020. Further interventions to support the liquidity of companies are also contained in Decree-Law No 34 of the 2020.

The economic support measures for businesses adopted with the decrees of March–May 2020 (Decree-Law No 18/2020, Decree-Law No 23/2020 and Decree-Law No 34/2020) are essentially attributable to the following main lines of intervention: liquidity support; export and internationalization support; capitalization support and non-repayable grants; suspension of certain obligations and tax payments, as well as temporary relief on the fixed costs of electricity bills for low-voltage non-domestic users; interventions for companies in crisis, industrial reconversion and development contracts; protection of the national economic and business fabric through changes, some of which are temporary, to the exercise of special powers in sectors of strategic importance (so-called golden power).

Among the measures for companies in crisis, industrial reconversion and development contracts, the following interventions are highly important.

Decree Law No. 18/2020 refinanced the measure of development contracts by €400 million for 2020 (Article 80). The Ministero dello Sviluppo Economico (MISE) Directive of April 15th, 2020 provided for the allocation of resources.

Finally, it should be noted that Law Decree no. 18/2020 authorized the Extraordinary Commissioner for the Epidemiological Emergency to provide funding to companies producing medical devices and personal protective equipment, using INVITALIA as the entity managing the measure. To this end, expenditure of EUR 50 million for 2020 has been authorized (Article 5). The aid scheme was authorized by the EU Commission (on 22nd March 2020). The Ordinance of the Extraordinary Commissioner of 23rd March 2020 (published in the Official Journal on 24 March 2020) implemented the measure.

The resources were assigned to the granting of aid to investment programs aimed at increasing the availability of medical devices and personal protection equipment in the national territory through the expansion of the capacity and/or the reconversion of an existing production unit. The facilities consist of subsidized financing of up to 75% of eligible expenditure. The maximum amount of the facilities that can be granted, in terms of aid (intended as Gross Grant Equivalent), may not exceed 800,000 euros, in accordance with the European Commission Communication of 19th March 2020 - COM (2020) 1863 final - "Temporary Framework for State aid measures to support the economy in the current COVID-19 outbreak".

Manufacturing masks, gowns, gels and disinfection products, plexiglass spacers, medical devices. These are some of the production reconversions following the pandemic of companies in most of the textile-fashion sector, but also plastics, chemicals, cosmetics, manufacturing, medical, graphics and printing [20].

For some sectors, textiles and chemicals, the new production is opening stable business opportunities in the post 2020 long time crisis, through new channels, which also open opportunities for professional integration.

More than two thirds of companies in the chemical sector, which in the emergency produced alcohol-based disinfectant gels for the hospital sector, are planning to permanently convert, but now intend to extend to direct sales to consumers.

And two thirds of the companies in the plastics sector, which have taken the opportunity to make plexiglass spacers to be installed in the companies, will not stop production. By virtue of a demand that is still expected to be sustained, moreover, more than half of the companies in the textile sector, which are now also aiming to create joint ventures with fashion companies, and almost all the companies in the print sector, which have activated new channels, will maintain active production of masks.

Not all companies, however, are planning to maintain the conversion once the normality is restored, with profound differences between sectors, due to the specificities of the productions.

These are mainly temporary reconversions, on the other hand, for fashion companies that have turned for a few weeks to the production of masks and gowns, as for those in the automotive, cosmetics, medical devices, and manufacturing sectors.

In addition to interventions aimed solely at conversion, the whole world of work has had to face the need to change and adapt to the new situation. Another example of resilience, together with the reconversion of the production of different companies, was the adoption of smart working.

There are data on the transition to remote working collected by Associazione Italiana Esperti Infrastrutture Critiche (AIIC) with the help of other companies.

It became clear that before the crisis and therefore until 2019 in companies 71% of employees did not even know what remote working was. During the pandemic 97% of people said they had been working remotely all the time and 43% of people interviewed said they would continue to work remotely.

Regarding the impact on the IT budget: 30% of companies said that investments on the 2020 roadmap projects reset and/or moved to 2021 or suspended.

In contrast, 30% of companies stated that investments will continue without any impact on the 2020 roadmap projects.

Finally, 60% of companies say they still do not know how to proceed with the investments.

The company management, however, has the advantage of being able to provide incentives for sanitization and safety at work: for companies are introduced incentives for sanitization and increased safety at work, through the granting of a tax credit equal to 50% of expenses up to a maximum of 20 thousand euros, and contributions through the establishment of an Inail fund.

The pandemic emergency has not only produced a strong acceleration of digital transformation, smart working and strong demands related to logistics, but also interesting productive reconversions, together with the consciousness of the complex interrelation through different sectors and their supply chain.

For SMEs, the introduction of new products has often meant a real revolution in the business, but able to ensure continuity in production that would otherwise have stopped. Moreover, in case the reconversions are expected to be permanent, are requiring new professional figures to support the activity.

And, most of all, the emergency confirmed the relevance of the human factor.

3. Concrete approaches to critical infrastructure protection

3.1 Supply chain continuity management and lack of manpower during the pandemic

Supply Chain Continuity Management (SCCM) must be considered as a necessary evolution of Business Continuity Management (BCM) models. SCCM is outlined in the ISO 22318 standard which is part of the group of standards for continuity management including ISO 22301, ISO 22313 Security and resilience (ISO 22318), and ISO 28000, which specifies the requirements for a security management system, including those aspects critical to security assurance of the supply chain. SCCM defines continuity in relation to external supplies, third parties or internal entities that play a supplier role in the context of the organization.

The simplified representation of the supply chain therefore provides a composite structure of internal and external suppliers (considering also the flexibility applicable to the relationships between the suppliers) that contribute to the operations of an organization and consequently of its customers.

If the relationship with suppliers is characterized by assets that are mainly intangible and movable and therefore related, for example, to the exchange of information or movable consumer goods, there will be greater control. An example in this sense, during the pandemic emergency management consisted in the possibility of maintaining relationships with suppliers through forms of smart working. This form of collaboration and coordination has been possible mainly between entities operating in sectors consisting of intangible assets such as professional, scientific and technical activities, financial and insurance activities, the activities of extraterritorial organizations, public administration and most professional services and, in general, all sectors that have not been affected by the suspension decrees.

In any case it will be necessary to have a management plan in case of crisis or incidents involving the supply chain.

The adoption of such measures will result in increasing control over the value chain in relation to an organization. In particular, the analysis carried out on the supply chain gives visibility to the mapping of interdependencies between different sectors allowing an analysis that goes beyond the single organization. Network analysis techniques could be combined with criticality and reliability metrics in order to produce composite methods that provide useful information to stakeholders [21].

As for ISO 22301, to plan the SCCM it will be necessary to carry out Impact Analysis activities with the individual suppliers involved, distinguishing critical suppliers from non-critical suppliers. For all relationships with critical suppliers, the guarantee of continuity can be determined by identifying a SCCM strategy to be agreed in transparency with these suppliers. Some strategic approaches may be:

- Reducing dependence on a supplier: direct engagement of substitute suppliers for a specific service; increasing on-site stock holding; establishing alternative solutions.

- Increasing resilience: loss mitigation; establishing mutual support policies with competitors.

- Working with suppliers: creating partnerships with suppliers; setting performance standard; monitoring and dealing with suppliers to increase their resilience; including SCCM requirements in supplier contracts.

The direct effects of the suspension decrees concerned the sectors directly involved and all those sectors that had to sustain the labor shortage caused by the lockdown. While other sectors not directly involved in the suspension decrees, such as financial services or wholesale trade, or sectors more prone to targeted reconversions and the adoption of smart working strategies such as online trade or the fashion sector, were able to stem the direct impact of the emergency or even profit from it.

The Italian National Institute of Statistics in May 2020 has provided a wide range of data and information about the positioning and contribution of the sectors within the Italian production system.

The database is based on the Extended Statistical Register on Economic Performance of Enterprises (Frame-SBS), which contains individual data on all industrial and service enterprises active in the country (about 4.4 million units), supplemented with additional statistical registers that provide detailed information on the characteristics of the employment, as well as import and export enterprises. The data have been further integrated with indicators taken from Italian Accounting.

Considering the enterprises that are part of the universe of reference of the system of Structural Business Statistics (SBS), those that from May 4 are operating in sectors still formally suspended are about 800 thousand (19.1% of the total), with an employment weight of 15.7% on the total of the sectors of industry and market services (excluding the financial sector) [22].

Figure 1.
Unavailability of manpower.

By revising and analyzing the Istat dataset updated in May 2020 [23] with regard to the pandemic, it can be observed in the **Figure 1** below that the unavailability of manpower has most directly affected the following sectors in percentage terms:

1. Other mining and quarrying activities; creative, artistic and entertainment activities; travel agency, tour operator and reservation services and related activities; libraries, archives, museums and other cultural activities; rental and operative leasing activities; real estate activities; activities concerning lotteries, betting, gambling houses; Sports, entertainment and leisure activities; construction of buildings; Mining of metal ores; Manufacture of other transport equipment; Manufacture of leather and related products; Manufacture of motor vehicles, trailers and semi-trailers; Manufacture of furniture; Tobacco industry; Metallurgy; Advertising and market research: 100%

2. Manufacture of clothing, manufacture of leather and fur articles: 98,48%

3. Manufacture of fabricated metal products (except machinery and equipment): 93,98%

4. Manufacture of other non-metallic mineral products: 92,85%

5. Food service activities: 90,91%

6. Manufacture of machinery and equipment NCA: 89,48%

7. Textile industries: 86,77%

8. Other personal service activities: 83,46%

9. Wholesale trade (except of motor vehicles and motorcycles): 67,23%

10. Manufacture of rubber and plastic products: 63,67%.

3.2 Approaches to supply chain what if analysis: dependencies trees

Considering the analyses and remediation plans structured to protect the SCC, it is possible to structure What If models oriented to predict the consequences linked to the lack of a supply.

In relation to the manpower issue, for example, it is possible to structure time-oriented models that consider the negative effects of the manpower.

The Domino Effect methodology applied to manpower aims to study and quantify the consequences of a negative event that causes a lack of personnel and/or supply chain. The model is configured as a visualization of the propagation over time of the negative effects caused by the unavailability of a certain percentage of company personnel.

Such a predictive model can allow the decision maker to simulate different crisis scenarios resulting from the loss of personnel based on the formal organizational structure of the company. In order for the model to be effective, however, it will be essential to feed the model and the collection of information starting from the analysis of the organizational chart and the company function chart.

Information is needed that can be traced back to the following organizational areas:

- Administration (ADM)
- Actors in charge of Crisis Management (CM)
- Functions that have relationships with critical suppliers (SUP)
- Business (BSS)
- Commercial (COM).

The holistic evolution of this model consists in describing the interdependencies between different sectors starting from the simulation of a disservice concerning a sector. The generic example below can be applied to a single reality in order to understand what long-term effects the lack of manpower, considered as a distinguished sector, could have on the operational continuity of the organization itself (**Figure 2**).

The severity of the dependency corresponds to the extent to which the Quality of Service (QoS) perceived by the user is deteriorated. Depending on the item, the degradation can be measured by the variation of some specific parameters (coverage, signal reception, delivery time, etc.) with respect to the normal QoS values. In general, the measures that allow to characterize the QoS can be traced back to the general concepts of availability and capacity: the quality with which the service is

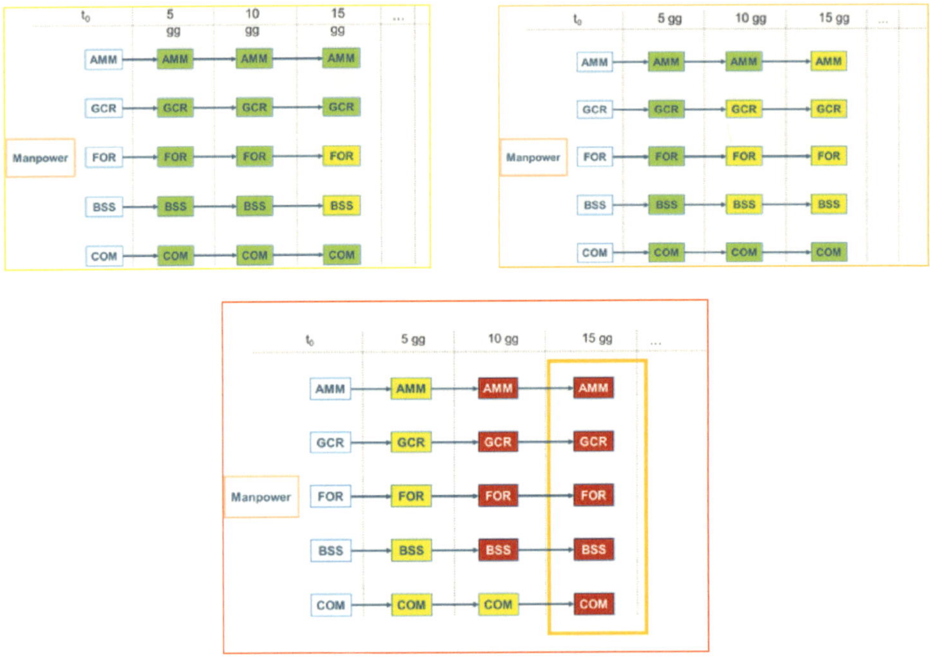

Figure 2.
Manpower cascading effect on organizational areas.

provided can be described by quantifying the quantity of items provided in comparison to the demand and the time in which the service is actually available. The choice of the temporal moments in which to sample the phenomenon varies according to the item represented.

Metrics commonly agreed to in these cases include: Abandonment Rate; ASA (Average Speed to Answer); TSF (Time Service Factor); FCR (First-Call Resolution); TAT (Turn-Around Time); TRT (total resolution time); MTTR (Mean Time To Recover).

Starting from the elaboration of matrices that consider dependency relations, to represent a domino effect map it is necessary to apply a "filter" based on the degradation level of the service. an item will be considered compromised (and therefore will be represented in the domino effect map) only if the QoS degradation will be higher than a certain threshold, so the service is not considered acceptable (outage).

Various methods are described in the literature to perform this assessment. In general, the most common approaches consist in identifying some indicators that describe the various aspects of the consequences caused by an out of service event.

These indicators can fall into the following categories:

- number of people (evaluated in terms of people impacted by the disruption)
- economic damage (assessed in terms of the extent of economic losses and/or deterioration of products or services)
- effects on public opinion (assessed in terms of impact on public confidence, physical suffering, and disruption of daily life).

Simulation of interdependencies and graph-based model to understand critical infrastructure interdependencies are proposed in literature [24–27].

The graphical output here proposed (**Figure 3**) from the described model consists of dependency trees, time-oriented, that describe the collapse of the internal structure of an organization following the manpower "sector" unavailability. This model can be applied to a single organization based on its SC analysis starting considering one or more products and services sectors.

By re-analyzing the ISTAT indices and considering the main sectors activated by the sectors impacted by the manpower shortage, it is possible to identify which related sectors have been most impacted by service interruptions than those listed above.

The sectors impacted indirectly by the shortage of manpower compared with the interruptions of those impacted directly are as follows:

- Rental and management of owned or leased properties
- Legal and accounting activities
- Road freight transport, removal and pipeline transport
- Financial service activities (except insurance and pension funding)
- Wholesale
- Manufacture of fabricated metal products (except machinery and equipment)

As we can see in **Figure 4**, some sectors such as Financial Services Activities that did not undergo significant effects during the first phase of the lockdown, are subject to an indirect impact due to the activity suspension of their main suppliers.

3.3 Augustus method and its application by the Italian civil protection

The Augustus method can be considered as another concrete approaches to Critical Infrastructure protection.

The Method is a tool used by the Civil Protection Department of the Italian Republic for emergency planning. The Augustus Method was created in order to equip the Italian Civil Protection Service with a unified strategy for planning the Civil Protection assistance at various levels of competence.

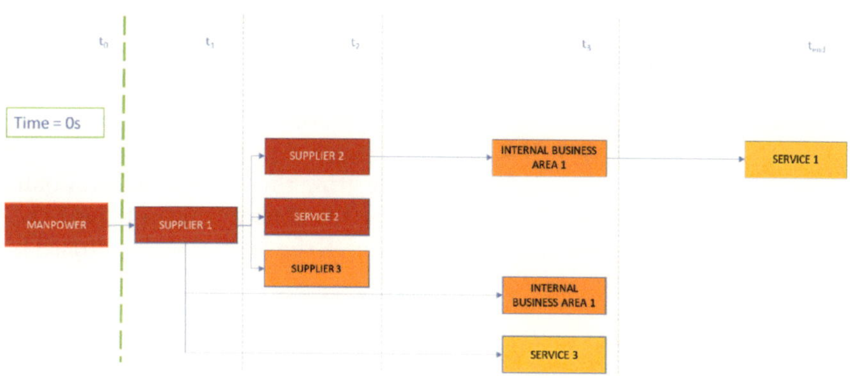

Figure 3.
Manpower dependency tree.

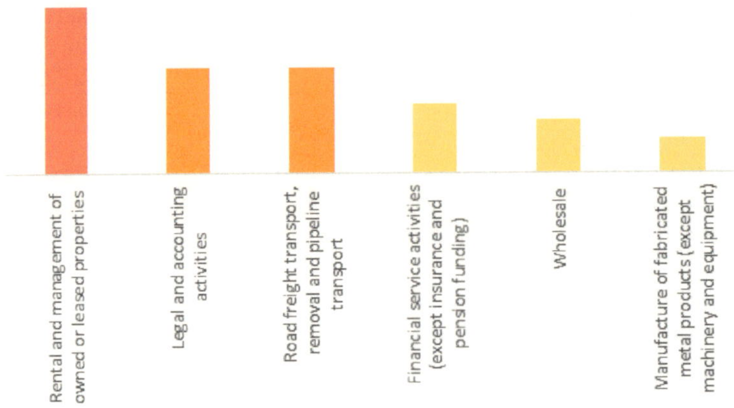

Figure 4.
Index of indirect impact on SCC for other sectors not highly affected by unavailability of manpower.

This method is named after the Roman Emperor Augustus (27 B.C. to 14 A.D.), who affirmed that: "*The value of planning decreases with the complexity of the state of things*." In detail, Augustus stated that it is impossible to plan a strategy in the smallest detail, because the event when it happens will always present in a different way. The Augustus Method is generated from the need to harmonize the directions of emergency planning.

This approach to the complexity of modern reality was structured and adapted by Elvezio Galanti, who considers the "emergency" (a public situation of particular difficulty and danger) an "organism" with its own life and composed by physiological functions (endocrine system, cardiology, etc.), each one specialized in its own field in which normally carries out its ordinary activity. In the context of civil protection, the "organism" is defined as the territory in which they normally act, and each one because of its specific functions (municipal, regional, health, transport, etc.). In the event of a disaster, these activities must all work together and in synergy.

The Augustus Methodology highlights, therefore, a fundamental aspect of the functioning of the Italian Civil Protection: its systemic nature. A complex apparatus made up of different elements and different organizations, resulting from the functioning of different systems in interaction with each other and with the other organizational systems [28].

In the preventive design phase, the Civil Protection, first of all, must work to collect information (time of occurrence of an event, geological conformation, productive fabric, urban fabric, etc.), then it must proceed with basic examinations (hazard analysis, vulnerability analysis, etc.) and finally a first diagnosis will be made (scenario, i.e. what I expect to happen) and for this reason, facilities will be arranged (monitoring networks, cleaning of riverbeds, seismic adaptation of structures, etc.).

In the absence or in the impossibility of activating these protocols, minimum measures of confrontation will be taken through the constitution of a "resilient cell" to manage the "big 5", i.e. five macro-areas in which the operational approach is divided into "acute emergency". These are:

1. identification of sites per control room;

2. entry points for expected rescue;

3. reception areas and first assistance to the population;

4. identification of proximity sites to coordinate local interventions;

5. assistance to the population (health and management of any temporary camps for reception and stay).

In the "acute" emergency scenario the Augustus Method becomes a good practice to manage the situation through the identification of 14 basic support functions, or support, that match all the competent and specific institutional figures for each function at territorial level and that contribute to its ordinary and extraordinary functioning. These functions are usually involved during the emergency itself, while in the study phases prior to the emergency, such as forecasting and prevention, they are deactivated and delivered to their specific and ordinary institutional functioning. These functions are: F 1 - Technology and planning; F 2 - Health, social and veterinary assistance; F 3 - Mass-media and information; F 4 - Volunteering; F 5 - Materials and means; F 6 - Transport, traffic and roads; F 7 - Telecommunications; F 8 - Essential services; F 9 - Census of damage to persons and property; F 10 - Operational facilities; F 11 - Local authorities; F 12 - Hazardous materials; F 13 - Assistance to the population; F 14 - Coordination of operational centres.

The design of all coordinated activities and procedures of Civil Protection to respond to any disaster event that is expected in a specific territory is called "Emergency Plan". The Emergency Plan must be implemented:

1. Forecasting and Prevention Programs

2. Information related to:

    - physical processes causing the risk conditions and their assessments

    - precursors

    - events

    - scenarios

    - available resources.

Therefore, it is necessary to represent graphically the information necessary for the characterization of possible risk scenarios for the implementation of intervention strategies for the rescue and management of the emergency, rationalizing and targeting the use of men and means.

According to the Method, the following conditions determine the success of a civil protection operation [29]:

- unitary direction: the unitary direction of emergency operations is implemented through the coordination of a complex system and not in a sectoral vision of the intervention.

- communication: constant exchange of information between the central and peripheral Civil Protection system.

- resources: rational and timely use of the resources really available and the availability of the men and means suitable for intervention.

The Emergency Plan structured according to the Augustus Method must be able to answer the following questions:

- what calamitous events may reasonably affect the municipality?
- which people, facilities and services will be affected or damaged?
- what operational organization is necessary to minimize the effects of the event with particular attention to the protection of human life?
- to whom are the different responsibilities at the various levels of command and control for emergency management assigned?

To satisfy these needs, it is first of all necessary to define the risk scenarios on the basis of the vulnerability of the portion of the territory concerned (areas, population involved, damaged structures, etc.) in order to have a global and reliable picture of the expected event and therefore to be able to dimension in advance the operational response necessary to overcome the disaster with particular attention to the protection of human life (how many firefighters, how many volunteers, which command and control structures, which roads or escape routes, which shelter structures, health areas, etc.).

The Emergency Plan is therefore a working tool calibrated on a likely situation based on scientific knowledge of the state of risk of the territory, which can be updated and integrated with reference to the list of men and means, but especially when new knowledge is acquired on the conditions of risk involving different assessments of the scenarios, or even when new or additional monitoring and warning systems to the population are available [30].

On the provincial level, the Emergency Plan will identify, at an inter-municipal or provincial scale: on the one side the situations that can configure a more extensive emergency of the single municipality, on the other side the situations, even localized, of greater risk, pointing out, when necessary, the need for an in-depth study of some aspects related to the Municipal scale.

On municipal level, a more detailed level of information is needed to allow the operators of the various components of the Civil Protection to have a reference framework corresponding to the size of the expected event, the population involved, the alternative road system, possible escape routes, waiting areas, shelter, storage areas and so on. Considering that the risk present in a given territory may refer to different types of events (floods, earthquakes, landslides, etc.), the Emergency Plan must provide for one or more "risk scenarios", which must or may correspond to different types of intervention.

The Italian Civil Protection assumes primary and decisive roles on the institutional scene of civil protection in Italy. This body sums up three fundamental structures at national level:

- the Civil Protection Department at the Presidency of the Council of Ministers
- the General Directorate of Civil Protection and Firefighting Services at the Ministry of the Interior
- the National Seismic Service at the Department of National Technical Services (currently dependent on the Ministry of Public Works).

The Civil Protection plays a key role in the management of national emergencies but not only: the possibility of being activated by the Prefect (Prefetto) for emergencies and in particular cases also for events at local level, makes the Civil Protection an entity that can operate de facto across the board. The Prefect is the cornerstone of the command and coordination structure of the civil protection operational system.

Another key player is represented by the Mayor. He is the determining element in the operational chain of civil protection at municipal level in the assumption of all responsibilities related to civil protection tasks: from the preventive organization of control and monitoring activities to the adoption of emergency measures aimed primarily at safeguarding human life.

It is appropriate, at this point, to make one final consideration: the Emergency Plan is drawn up in any case on the basis of the scientific knowledge possessed at the time of writing, without waiting for studies in progress or future assignments or improvements. An "expeditious" plan, even if imprecise and precautionary, is better than no plan at all. As soon as possible, the Emergency Plan will be reviewed, improved, and completed with more data and more scientific bases.

The key concept of contingency planning is to try to predict all possible variables, however, it is necessary to be aware that it will always be possible, in any emergency, to face something unforeseen.

3.4 The Italian civil protection strategy for the management of the 2020 crisis

The coordination of the members of the National Service of Civil Protection is happening according to the provisions of the Augustus Method thanks to the synchronism of the representatives of each operational function (Health, Volunteering, Telecommunications, etc..) to interact directly with each other.

The intervention model adopted by civil protection for the management of the epidemiological emergency [31] based on the definition of the chain of command and control, the communication flow and the procedures to be activated in relation to the emergency state determined by the spread of the pandemic.

The chain of command and control includes the following levels of coordination:

- National level: the Head of the Civil Protection Department ensures the coordination of the necessary interventions, making use of the Department, the components, and operational structures of the National Civil Protection Service, as well as implementing entities. At the Department of Civil Protection is active the Civil Protection Operational Committee, with the task of ensuring the contribution and support of the National Civil Protection System on the basis of the health indications defined by the Ministry of Health, which makes use of the ISS (Istituto Superiore Sanità) and the Scientific Technical Committee specifically established with the OCDPC 630/2020 at the Department.

- Regional level: at all Regions must be activated a regional crisis unit, which operates in close connection with the SOR - Regional Operations Room, which must provide for the participation of the Regional Health Contact, which operates in connection with the Health Director of the local health agencies, and in constant contact with a representative of the Chief Prefecture, in order to ensure the connection with the other Prefectures - UTG of the regional territory.

- Provincial level: in the provinces in which at least one person is positive for whom the source of transmission is unknown or in any case where there is a case not attributable to a person from an area already affected by the virus, as provided by art. 1, paragraph 1 of Decree-Law no. 6 of 23.02.2020, the Prefect or his delegate provides for the activation of the CCS - Rescue Coordination Centre

- Municipal level: in the municipalities or areas in which at least one person is positive for whom the source of transmission is unknown or in any case where there is a case not attributable to a person from an area already affected by the aforementioned virus, as provided by art. 1 paragraph 1 of Decree-Law no. 6 of 23.02.2020, the Mayor or his delegate provides for the activation of the Municipal Operations Centre - COC of the municipality involved and neighboring municipalities in order to implement possible preventive actions.

Therefore, in order to cope with the pandemic and in accordance with the provisions of the Augustus Method, collaborative decision-making processes have been initiated in real time in the operational rooms of the various levels such as:

- Centro Coordinamento dei Soccorsi (CCS) - Rescue Coordination Centre
- Centro Operativo Comunale (COC) - Municipal Operations Centre
- Centro Operativo Misto (COM) - Mixed Operations Centre.

The CCS is the main body at provincial level and is chaired by the Prefect or his delegate. By COC is meant the Municipal Operations Centre, responsible for the activities at municipal-local level, whose maximum point of reference is the Mayor or his delegate (Law 225/1992 - Art. 15). Finally, the COM is the Mixed Operations Centre. They can be more than one and set up ad hoc to be as close as possible to the place of the event.

Originally established as emergency operational centres (i.e. support and operational coordination structures set up and organized exclusively in the full management phase of the emergency following catastrophic events), over time the term has moved to a broader interpretation of the term which also involves structures and organizational divisions of one or more local administrations in the construction of the local civil protection system as well as emergency planning activities to be carried out in ordinary time.

In this emergency caused by the pandemic, a key role is played by the COC, which have been activated in many Italian municipalities [32].

Specifically, the Mayor makes use of the COC to ensure the direction and coordination of rescue and assistance services to the population within his municipal territory in relation to the declaration of the state of emergency issued by the Italian Government. The choice of the location of this Centre must be in earthquake-proof structures, in areas with easy access and not vulnerable to any kind of risk. These facilities must be equipped with a square of enough size to accommodate heavy vehicles and anything else needed in a state of emergency. The COC is responsible for the decision-making levels of the entire municipal structure, summarized in the trade union responsibilities referred to in the previous paragraphs; as a rule, the decision-making level is taken by the Mayor who, through a municipal civil protection system, identifies the actions and strategies necessary to try to keep the infection curve and morbidity index under control. The COC operates in a place of coordination called "operations room" where all the news related to the event converge and where decisions are taken to overcome it. In many municipalities, the COC has been activated by the Mayor as an immediate consequence of the increase in infections within the national territory, and not necessarily in the municipal one, and it will remain operational until the resolution of the pandemic crisis [33].

According to the Civil Protection Operational Measures for the management of the epidemiological emergency [31] actions and operational measures

DOI: http://dx.doi.org/10.5772/intechopen.94894

identified for each level of coordination, without prejudice to the provisions issued by the Ministry of Health, are as follows:

- information to the population
- activation of local volunteering, in connection with the levels of coordination above
- organization of actions at the municipal level, in connection with the regional and provincial level, actions to ensure the continuity of essential services, as well as the collection of waste in areas affected, or that may be affected, by urgent measures of containment
- organization of actions at the municipal level, in connection with what has been prepared at the regional level, actions aimed at ensuring the continuity of the supply of basic necessities (including fuel supplies) in the areas concerned, or that could be affected by urgent containment measures;
- planning, or possible activation, of the actions of assistance to the population of the municipalities concerned, or that could be affected by urgent containment measures
- planning and organization of home care services for persons in home quarantine (e.g., basic necessities, medicines, pre-packaged meals...), possibly carried out by personnel of volunteer organizations, appropriately trained.

At this point, it can be stated that the success of a civil protection operation can be achieved if three parameters are satisfied: coordination, communication, and resource management.

4. Conclusion

As with any crisis management strategy, resilience strategies must be planned and prepared during the "peace" period and then implemented, appropriately adapted, during crisis situations. The variable structure, and a proactive response, is what succeeds in giving us a continuity and dealing successfully with the complexity.

Labor shortages directly affected all those sectors that had to close due to the impossibility to convert their business using smart working. Some activities, although part of sectors not directly involved in the lockdown, were indirectly affected by labor shortages caused by the inability of seasonal and commuting staff to move. Finally, the indirect repercussions that have affected those activities that, while remaining operational, have suffered significant economic repercussions due to the interruption of their supply chain caused by the shortage of labor in other sectors.

To be considered in the degree of dependence that an organization might have on its suppliers, beyond its intrinsic resilience, is the degree of flexibility applicable to relations with the various suppliers.

To plan the SCCM it will be necessary to carry out Impact Analysis activities with the individual suppliers involved, distinguishing critical suppliers from non-critical suppliers. For all relationships with critical suppliers continuity can be determined by identifying a SCCM strategy to be agreed transparently with these suppliers. Some strategic approaches may be:

- Reducing dependence on a supplier: direct engagement of alternative suppliers for a given service; increasing on-site stock holding; establishing alternative solutions.

- Increased resilience: mitigation of losses; identification of a set of alternative suppliers; establishing mutual support policies with competitors.

- Working with suppliers: creating partnerships with suppliers; setting performance standards (including through SLAs); monitoring and dealing with suppliers to increase their resilience; including SCCM requirements in supplier contracts.

The adoption of these measures will result in increasing control over the value chain in relation to an organization. In particular, the analysis carried out on the supplier chain gives visibility to the mapping of the interdependencies between the different sectors enabling an analysis that goes beyond the single organization.

Therefore, maximum flexibility and, at the same time, the ability to create the preconditions (e.g. through exercises) is needed to ensure that the best conditions for success are in place in these cases as well.

Moreover, most of all, the 2020 crisis confirmed the relevance of the human factor.

The Italian case is an example of how the set of private initiatives, the support of adequate policies of incentives and support from the State, together with a strong sense of solidarity with the population, can represent a positive reaction to a negative event, and that business strategies oriented towards business continuity are the basis for the development of resilience in the productive sector, and the resilience of the Critical Infrastructures.

Author details

Luisa Franchina[1], Alessandro Calabrese[2], Enrico Scatto[2] and Giulia Inzerilli[2]*

1 AIIC, Rome, Italy

2 Hermes Bay, Rome, Italy

*Address all correspondence to: g.inzerilli@hermesbay.com

IntechOpen

© 2020 The Author(s). Licensee IntechOpen. This chapter is distributed under the terms of the Creative Commons Attribution License (http://creativecommons.org/licenses/by/3.0), which permits unrestricted use, distribution, and reproduction in any medium, provided the original work is properly cited.

References

[1] P. Hokstad, I. Utne, I. Vatn; Risk and Interdependencies in Critical Infrastructures: A Guideline for Analysis; Springer; 2012

[2] Y. Deng, *L. song*, Z. Zhou, P. Liu; Complexity and Vulnerability Analysis of Critical Infrastructures: A Methodological Approach; Mathematical Problems in Engineering. 2017

[3] M. Tvaronaviciene; Towards internationally tuned approach towards critical infrastructure protection; Journal of Security and Sustainability Issues; 2018

[4] Gorzeń-Mitka, M. Okręglicka; Managing Complexity: A Discussion of Current Strategies and Approaches; Procedia Economics and Finance. 27. 438-444; 2015

[5] I. Sperstad, G. Kjølle, O. Gjerde; A Comprehensive Framework for Vulnerability Analysis of Extraordinary Events in Power Systems; Reliability Engineering; 2019

[6] R. Napoli; Coronavirus e infrastruttura elettrica; AIIC Newsletter no. 06/2020

[7] R. Setola; La sicurezza nazionale alla prova della resilienza. L'analisi di Setola; Formiche, 1st April 2020

[8] Principi Precauzionali Per Gli Operatori Di Infrastrutture Critiche ai fini della continuità in sicurezza del servizio di interesse pubblico; Ufficio del Consigliere Militare, Segreteria Infrastrutture Critiche; 26th March 2020

[9] COVID-19: assicurata la piena funzionalità del Centro di Controllo Galileo al Fucino; Telespazio, 29th April 2020

[10] Chiappetta; Italia next ovvero un decalogo per il dopo Coronavirus. Intervento di Chiappetta; Formiche, 5th April 2020

[11] E. Luiijf, A. Nieuwenhuijs, M. Klaver, M. Eeten, E. Cruz; Empirical findings on European critical infrastructure dependencies; International Journal of System of Systems Engineering; 2. 3. 10.1504/IJSSE.2010.035378; 2010

[12] G. Pisano, R. Sadun, M. Zanini; Lessons from Italy's Response to Coronavirus; Harvard Business Review; 27th March 2020

[13] Decreto Presidente del Consiglio dei Ministri 4th May 2020

[14] Emergenza Covid, oltre 3 mila ventilatori polmonari dalla sinergia tra Fca e Siare, la Repubblica; 9th July, 2020

[15] È genovese, il brevetto che converte le maschere da sub in protezioni contro la Covid-19; Il Secolo XIX; 24th March, 2020

[16] *A. Carli*; Mascherine e respiratori, ecco le fabbriche che si riconvertono; il Sole 24 Ore, 24th March, 2020

[17] P. di Lazzaro; Le mascherine della ex Mabro; Rainews; 17th March 2020

[18] A. Nasso; Coronavirus, Roncato ferma produzione valigie: "Facciamo mascherine, nel nostro futuro anche guanti e visiere"; la Repubblica; 28th April, 2020

[19] European Commission Press Release; State aid: Commission approves €50 million Italian support scheme for production and supply of medical equipment and masks during Coronavirus outbreak; 22nd March 2020

[20] Newsroom; Face masks, sanitiser gel and ventilators – those Italian factories switching their production; Mornong Future; 4th May 2020

[21] J. Williams; Critical Flow Centrality Measures on Interdependent Networks with Time-Varying Demands; University of Toronto, Department of Computer Science, Canada; 2019

[22] ISTAT; Dataset contributo e posizionamento dei settori produttivi; 12nd May 2020; Available from: https://www.istat.it/it/archivio/239854

[23] ISTAT; Nota esplicativa situazione 4 maggio; 11st May 2020; Available from: https://www.istat.it/it/files//2020/04/nota-esplicativa_situazione_04maggio.pdf

[24] N. Svendsen, S. Wolthusen; Graph Models of Critical Infrastructure Interdependencies; First International Conference on Autonomous Infrastructure, Management and Security, AIMS 2007

[25] S. Dominique, Y. Barbarin, M. Eid; Preparing for the Domino Effect in Crisis Situation D2.1 State of the Art of the R&D Activities in Cascade Effect & Resilience and Global Modelling; Report PREDICT-20151217-D2-1V3; 2018

[26] O. Pala, P. Schrum; Simulating Infrastructure Outages: An Open-Source Geospatial Approach; Conference: GeoInformation for Disaster ManagementAt: Istanbul Technical University; 2018

[27] A. Rahman, Ḥāfiẓ; Modelling and simulation of interdependencies between the communication and information technology infrastructure and other critical infrastructures; 2009

[28] E. Galanti; Il Metodo Augustus; DPC INFORMA "Periodico informativo del Dipartimento della Protezione Civile" (year II; number 4); 1997

[29] Presidency of the Council of Ministers, Italian Civil Protection Department; National Risk Assessment; December 2018

[30] Provicial Emergency Plan; Metodo Augustus e Funzioni di Supporto; Assessorato alla Protezione Civile; 2008

[31] Dipartimento di Protezione Civile; Misure operative di protezione civile per la gestione dell'emergenza epidemiologica da Covid-19; 4th May 2020

[32] Comune di Olevano sul Tusciano; Istituzione del "Centro Operativo Comunale (COC)" per l'emergenza di protezione civile COVID-19, 23rd March 2020; https://www.comune.bergamo.it/node/188215

[33] Emergenza Coronavirus Covid-19 Sul Territorio Nazionale – Attivazione C.O.C. 8th March 2020, https://www.olevanosultusciano.gov.it/avvisi/emergenza-coronavirus-covid-19-sul-territorio-nazionale-attivazione-c-o-c-centro-operativo-comunale/